# Networks: The Creation and Circulation of Knowledge from Franklin to Facebook

# Networks: The Creation and Circulation of Knowledge from Franklin to Facebook

### Edited by the
### American Philosophical Society

American Philosophical Society Press

Philadelphia

Transactions of the
American Philosophical Society
Held at Philadelphia
for Promoting Useful Knowledge
Volume 111, Part 4

ISBN: 978-1-60618-114-0
Ebook ISBN: 978-1-60618-119-5
U.S. ISSN: 0065-9746

**Library of Congress Cataloging-in-Publication Data**

Names: American Philosophical Society, editor.
Title: Networks : the creation and circulation of knowledge from Franklin to
    Facebook / edited by the American Philosophical Society.
Description: Philadelphia : American Philosophical Society Press, [2022] | Series:
    Transactions of the American Philosophical Society, 0065-9746 ; volume 111,
    part 4 | Includes bibliographical references and index. | Summary: "This
    volume considers historical networks of knowledge creation and dissemination
    in early America"— Provided by publisher.
Identifiers: LCCN 2022013010 (print) | LCCN 2022013011 (ebook) | ISBN
    9781606181140 (paperback) | ISBN 9781606181195 (mobi) Subjects: LCSH:
    Learning and scholarship—United States—History. | Science—United
    States—History. | United States—Intellectual life—1783-1865.
Classification: LCC AZ503 .N48 2022 (print) | LCC AZ503 (ebook) | DDC 001.2—
    dc23/eng/20220517
LC record available at https://lccn.loc.gov/2022013010
LC ebook record available at https://lccn.loc.gov/2022013011

Cover design by Eugenia B. González.

# Contents

# Contributors

**Lea Beiermann, PhD candidate,** Maastricht University, Maastricht, Netherlands

**Eileen Ka-May Cheng, PhD,** Sara Yates Exley Chair in Teaching Excellence, Sarah Lawrence College, Bronxville, New York

**Alicia DeMaio, PhD,** Upper Division History Teacher, Horace Mann School, New York, New York

**Peter C. Messer, PhD,** Associate Professor, History, Mississippi State University, Starkville, Mississippi

**George D. Oberle III, PhD,** History Librarian, Assistant Professor, George Mason University, Fairfax, Virginia

**Patrick Spero, PhD,** Librarian and Director, Library & Museum of the American Philosophical Society, Philadelphia, Pennsylvania

# Librarian's Note

---

Since its founding in 1743, networks have always been at the core of the American Philosophical Society. As the first learned society in North America, the APS was meant to connect leading Colonial thinkers, dispersed across the Eastern seaboard, together in one large correspondence network. The idea was that by sharing their research with each other, they would advance their collective knowledge, and in turn their respective communities. The APS was also meant to serve as the central organizing node of scholars in North America by forming an intellectual bridge upon which colonists could connect with their European peers across the Atlantic. In time, as the APS Membership became more active, it found news ways to leverage this network, including calling on Members to contribute important items to its burgeoning library. These traditions continue today. The APS Membership now numbers around 1,000, with 850 Resident Members in the United States and 150 International Members. The APS Library houses more than 14 million pages of manuscripts, many having come from Members and their colleagues. The Society hosts regular meetings of Members to cultivate new connections and showcase cutting-edge research.

More recently, the digital age has created opportunities for the APS to serve its mission of advancing knowledge and to think about its history, collections, and Membership. In 2017, the Society formed its Center for Digital Scholarship (CDS) with the mission to digitize materials, to serve scholars and researchers through the development of new projects and fellowships supporting work in the digital humanities, and through the creation of new discovery tools that allow more opportunities for users to explore the Library & Museum's holdings. The formation of the CDS has also enabled Library & Museum staff to develop open datasets based on material from the collections that are available to anyone who might like to share, reuse, remix, and evaluate them, through digital projects as well as in the classroom.

It was with this spirit in mind that the CDS turned its attention to one of the Society's founders and its first President, Benjamin Franklin, when they sought to launch a project that explored networks. After realizing that Franklin's ledger and account books sat largely underutilized by the scholars, the CDS began a multiyear project to transcribe and digitize the postal ledgers kept by Franklin during his time serving as Postmaster of Philadelphia from 1748 to 1752. The volume contains 14,756 entries from approximately 3,500 unique mail recipients across fifteen Colonial American cities. This data has been published by the *Magazine of Early American Datasets* and is publicly available in the APS Github repository, measures that ensure the availability and accessibility of this data to scholars worldwide.

The visualizations and other analysis generated from this data offer glimpses into daily life and early business ventures in Colonial Philadelphia, while also revealing some new and surprising insights. For example, an assessment of which residents received the most mail during the period in question confirmed that John Mifflin, a successful merchant, held the honor with 708 letters. More curious, however, was the second-highest recipient of mail—a man named William Vanderspiegel, who is largely unknown. Such revelations suggest the potential of digital datasets to raise new questions for further investigation and analysis, ones that scholars may have otherwise missed. The digital project also asked questions about the seasonality of correspondence, the finance behind it, and the postal connections Philadelphia has with different cities and towns.

But the APS has always been focused on the present as much as the past, and, although its postal ledger's project looked to the Colonial era, it also simultaneously supported a similar project focused on late-twentieth-century science. As part of "The Cybernetics Thought Collective," a web-portal hosted by The University of Illinois Archives, the APS also digitized materials that helped document the development of the multidisciplinary field of cybernetics. Its contributions included the personal papers and correspondence of Warren S. McCulloch, a pioneer in early computing and mathematical modeling who became the first elected president of the American Society for Cybernetics in 1964.

Inspired by these two projects, the Library & Museum decided to focus its 2019 spring symposium on the topic of networks. Although Colonial postal records and mid-twentith-century cybernetics seems an unlikely pairing for a symposium, it proved the perfect framework for a diverse set of papers and discussions that explored the ways that social, scientific, and intellectual networks have influenced and shaped the pursuit of "useful knowledge" from the eighteenth century through the present day.

The chapters that follow represent only a fraction of the papers presented during the June 2019 event; nonetheless, they successfully convey the wide breadth of topics covered. With subjects ranging from the APS's influence on the politics of knowledge in the Early Republic to the role of local correspondence networks in cultivating knowledge of natural history to the creation of scientific communities through the circulation of microscopic images, this volume shows that much still remains to be said about how networks create and circulate knowledge—and how knowledge is shaped by them.

*Patrick Spero*
Librarian and Director
Library & Museum of the
American Philosophical Society

# 1

# Plagiarism as Dialogue: George Chalmers, William Robertson, and John Marshall's *Life of George Washington*

## Eileen Ka-May Cheng

A widespread practice among Revolutionary and early national American historians, plagiarism has often been seen as a sign of the derivative and unscientific character of their scholarship.[1] Yet plagiarism represented an interpretive choice for these historians, enabling them to make themselves part of an imagined community of scholars. Serving as their equivalent of a retweet, or of a footnote in the scholarly realm, plagiarism provided them with a means of both disseminating and criticizing the information they gathered from other historians. This chapter examines the role of plagiarism in creating transatlantic networks of intellectual exchange by focusing on John Marshall's extensive plagiarism from the loyalist historian George Chalmers's *Political Annals* and the Scottish historian William Robertson's *History of America* for his history of the colonies.[2] Marshall was by no means alone in his plagiarism from loyalist and British histories, making his treatment of Chalmers and Robertson a valuable window into the function this practice served for early national historians. Abiel Holmes likewise plagiarized from both Chalmers and Robertson, as did David Ramsay. Among the other historians who plagiarized from Chalmers were Hugh Williamson and George Bancroft, while Ramsay, Hugh McCall, and William Gilmore Simms all plagiarized from the loyalist Alexander Hewatt's *An Historical Account of the Rise and Progress of the Colonies of South Carolina and Georgia.*[3]

Better known for his importance as chief justice of the Supreme Court, Marshall was also a serious historian who published a five-volume biography of George Washington in 1804. The first volume of the biography consisted of a general history of the colonies, which plagiarized from many different historians besides Chalmers and Robertson.[4] But Marshall singled out the

---

[1] On how the prevalence of plagiarism reflected the backward and unscientific state of historical scholarship in this period, see Orin G. Libby, "Some Pseudo–Historians of the American Revolution," *Transactions of the Wisconsin Academy of Sciences, Arts, and Letters* 13 (1900): 419–25; Sydney G. Fisher, "The Legendary and Myth-Making Process in Histories of the American Revolution," *Proceedings of the American Philosophical Society* 51 (April/June 1912): 56.

[2] On Marshall's plagiarism, see William A. Foran, "John Marshall as a Historian," *American Historical Review* 43, no. 1 (1937): 51–64.

[3] On the plagiarism of Chalmers, see Richard Vitzthum, *The American Compromise: Theme and Method in the Histories of Bancroft, Parkman, and Adams* (Norman: University of Oklahoma Press, 1974), 51–53, 62–65; and Eileen Ka-May Cheng, "Plagiarism in Pursuit of Truth: George Chalmers and the Patriotic Legacy of Loyalist History," in *Remembering the Revolution: Memory, History, and Nation Making from Independence to the Civil War*, eds. Michael A. McDonnell et al. (Amherst: University of Massachusetts Press, 2013), 144–61. On the plagiarism of Hewatt, see Elmer Johnson, "Alexander Hewat: South Carolina's First Historian," *Journal of Southern History* 20 (1954): 57–59; and Eileen Ka-May Cheng, "Plagiarism and the Nationalist Uses of Loyalist History: Alexander Hewatt and David Ramsay," in *The Consequences of Loyalism: Essays in Honor of Robert Calhoon*, ed. Rebecca Brannon and Joseph Moore (Columbia: University of South Carolina Press, 2019), 208–27.

[4] On Marshall as a historian, see Foran, "John Marshall," 51–64; Daniel Gilbert, "John Marshall and the Development of National History," in *Historians of Nature and Man's Nature: Early Nationalist Historians*, ed. Lawrence H. Leder (New York: Harper & Row, 1973), 177–99.

works of Chalmers and Robertson for their value as contributions to American history and relied particularly heavily on them. Praising Chalmers's *Annals* as a "very valuable work," Marshall likewise commended Robertson's history of New England and Virginia as "an elegant and valuable history" and openly admitted that he had "made large extracts" from it.[5] Writing at a time when historians still adhered to an editorial conception of authorship in which their purpose was to compile and synthesize the words and knowledge of their predecessors rather than to create a new and unique work of their own, Marshall saw nothing wrong with copying from his sources without providing quotation marks or proper attribution.[6] He even confessed to what would be considered plagiarism today when he acknowledged in his preface that "their very language has sometimes been employed without distinguishing the passages" by quotation marks. Yet he also recognized plagiarism as an offense when he expressed his hope that this acknowledgment would absolve him "of wishing, by a concealed plagiarism" to pass off the work of others as his own. Hence he did cite his sources at the end of each chapter in the 1804 edition of the first volume of his history, and provided more specific citations in the 1805 and 1824 editions of this work, although he continued to copy the language of his sources without quotation marks and did not always cite the sources he used.[7]

Although plagiarism did exist as a concept in Marshall's time, its meaning was contested and fluid, resulting in both its prevalence as a practice and the widespread accusations of plagiarism that marked the literary culture of this period.[8] For Marshall, plagiarism was a matter of hiding his reliance on other sources and claiming their research as his own, not using their language, as what was important for him was the information he obtained from them, not how it was expressed. I therefore use the term *plagiarism* as a shorthand to describe his appropriations and to dramatize the extent of his reliance on Chalmers and Robertson, even though he did not consider this plagiarism himself. In highlighting Mar-

[5] John Marshall, *The Life of George Washington* (Philadelphia: C. P. Wayne, 1805), 1:xiii.

[6] See Jay Fliegelman, *Declaring Independence: Jefferson, Natural Language, and the Culture of Performance* (Stanford, CA: Stanford University Press, 1993), 164–81, on the relationship between plagiarism and this editorial conception of authorship. On its implications for ideas about the historian's role in early national historiography, see Eileen Ka-May Cheng, *The Plain and Noble Garb of Truth: Nationalism and Impartiality in American Historical Writing, 1784–1860* (Athens: University of Georgia Press, 2008), 106–20.

[7] Marshall, *The Life of George Washington*, 1:x. For this reason, I refer to the 1805 edition of his history throughout the chapter. On Marshall's ambivalence about plagiarism, see Cheng, *Plain and Noble Garb*, 109–11.

[8] On these conflicts, see, for example, Fliegelman, *Declaring Independence*, 164–70; and Cheng, *Plain and Noble Garb*, 106–15.

shall's plagiarism, my goal is not to impugn him for intellectual dishonesty, but to historicize the concept by illuminating its function as a mode of communication and intellectual exchange. I thereby depart from the way that plagiarism has been treated in conventional accounts of American historiography as, at best, a sign of intellectual deficiency, or, at worst, a moral defalcation.[9] Instead, following the lead of scholars of British and continental European literary and historical culture, who have demonstrated the complex meanings of plagiarism for early modern and nineteenth-century writers, I show how the use of plagiarism as an interpretive tool and a means of producing knowledge intersected with Marshall's national-ist purposes.[10]

But how did a staunch nationalist like Marshall reconcile his plagia-rism from Chalmers's scathingly critical portrayal of the colonies with his celebratory view of the American past? Marshall did not copy uncritically from Chalmers but modified his appropriations to vindicate the colonists from Chalmers's attacks and justify the American Revolution. Plagiarism thus allowed Marshall to engage in a dialogue with Chalmers in which he expressed both his agreements and disagreements with his loyalist oppo-nent. The dialogue between Marshall and Chalmers was not just one way, for it was part of a multivocal conversation with Robertson. Robertson had himself relied heavily on Chalmers for his history of the British colonies in North America. Hence, frequently citing Robertson and Chalmers in tandem with one another, Marshall imbibed Chalmers's influence both directly and at one remove, through his influence on Robertson.

Yet, like Marshall, Robertson did not uncritically follow Chalmers, but reframed the material he took from him to further his own quest for impartiality. By rendering Chalmers's interpretation of Colonial history into a form that was more amenable to Marshall's nationalist purposes,

---

[9] See Libby, "Some Pseudo–Historians of the American Revolution," 419–25; and Orin G. Libby, "Ramsay as a Plagiarist," *American Historical Review* 7 (1902): 697–703, for the origins of this critical view, which has then been perpetuated by later accounts of American historiography, such as Harvey Wish, *The American Historian: A Social–Intellectual History of the Writing of the American Past* (New York: Oxford University Press, 1960), 41. For a more recent attack on how plagiarism contributed to the falsification of American history, see Peter Hoffer, *Past Imperfect: Facts, Fictions, Fraud–American History from Bancroft and Parkman to Ambrose, Bellesiles, Ellis, and Goodwin* (New York: Public Affairs, 2007), especially 13–25. For a forceful condemnation of plagiarism as a moral offense, see also Christopher Ricks, "Plagiarism," in *Plagiarism in Early Modern England*, ed. Paulina Kewes (New York: Palgrave Macmillan, 2003), 21–40.

[10] See, for example, Kewes, ed., *Plagiarism in Early Modern England*; Giovanna Ceserani, "Narrative, Interpretation, and Plagiarism in Mr. Robertson's 1778 'History of Ancient Greece,'" *Journal of the History of Ideas* 66, no. 3 (2005): 413–36; Tilar J. Mazzeo, *Plagiarism and Literary Property in the Romantic Period* (Philadelphia: University of Pennsylvania Press, 2006); Reginald McGinnis, ed., *Originality and Intellectual Property in the French and English Enlightenment* (New York: Routledge, 2009); and Paulina Kewes, *Authorship and Appropriation: Writing for the Stage in England, 1660–1710* (Oxford: Clarendon Press, 1998).

Robertson's work both served as a conduit to Chalmers's account and provided Marshall with a counter to it. Conversely, at other points, Marshall used Chalmers's history to counter Robertson's interpretation. Thus, in his plagiarism of both these historians, Marshall at once engaged in a direct dialogue with Chalmers and Robertson, and spoke to each of them through the other. This dialogue was not just about how to interpret the Colonial past, but also about the proper form in which to present that history, as Marshall divested his appropriations from Chalmers and Robertson of their broader philosophical element. By converting his account of the American colonies from an extension of philosophical history, as it was for Chalmers and Robertson, into a specifically American narrative, Marshall portrayed the developments he described as unique to America and thereby furthered his exceptionalist vision of American origins. Basing that vision on America's capacity to balance order with commercial growth, Marshall used his appropriations from Chalmers and Robertson to justify Colonial expansion, while at the same time expressing and assuaging his qualms about the costs of that growth for Native Americans and slaves.

A prolific writer who published on a wide variety of literary and historical subjects, Chalmers was best known to American historians for his works on Colonial and Revolutionary American history, his *Political Annals* (1780) and his *Introduction to the History of the Revolt of the Colonies* (1782). A lawyer by profession, Chalmers emigrated to Maryland from Scotland in 1763, but had to leave the colony in 1775 because of his loyalist activities.[11] A staunch defender of British authority, Chalmers lambasted the American colonists in his *Political Annals*, which covered their history to 1688, dating the roots of independence back to the very inception of the colonies. Structuring his narrative around the contrast between the loyalty and respect for tradition that he believed characterized Virginia and Maryland and the spirit of independence and innovation that defined the New England colonies, Chalmers attributed the Revolution to the New England colonists' fanatical opposition to British authority, which the weakness and vacillation of British policy makers allowed to infect the rest of the colonies and eventually resulted in American independence.[12]

---

[11] For background on Chalmers, see Grace Amelia Cockroft, *The Public Life of George Chalmers* (New York: Columbia University Press, 1933); Lawrence Henry Gipson, "George Chalmers and the *Political Annals*," in *The Colonial Legacy: Loyalist Historians*, ed. Lawrence H. Leder (New York: Harper Torchbooks, 1971), 13–14; John A. Schutz, "George Chalmers and *An Introduction to the History of the Revolt*," in *Loyalist Historians*, ed. Lawrence H. Leder (New York: Harper Torchbooks, 1971), 50–51.

[12] On Chalmers's interpretation of Colonial history, see Cockroft, *The Public Life of George Chalmers*, 59; Schutz, "George Chalmers," 58; Peter Hoffer, "Fettered Loyalism: A Re-Evaluation of Robert Proud's and George Chalmers' Unfinished Colonial Histories," *Maryland Historical Magazine* 68 (1973): 160–72; and Wesley Frank Craven, *The Legend of the Founding Fathers* (New York: New York University Press, 1956), 51–53.

At the same time, he blended his reverence for order and tradition with a dynamic vision of commercial progress derived from Scottish Enlightenment ideals.[13] Even as he condemned the disorder and rebelliousness of the New England colonists, he also admired the vigor and energy that enabled them to prosper economically. For Chalmers, the very liberty that made the colonists so fractious was also necessary for the economic prosperity that he identified with social progress and refinement, and he struggled to reconcile the tension between his desire for order and his faith in commercial progress. Firmly convinced that "[f]reedom then was the enlivening principle of their pursuits," Chalmers revealed how intractable the tension was between that freedom, with the prosperity it engendered, and social order when he concluded that "no man has been found who can unravel the intricate knot, by making the advantages of union and the interests of freedom coalesce."[14]

Chalmers's debt to the Scottish Enlightenment helps explain the appeal of his work for Robertson. A minister and leader of the Church of Scotland, Robertson made his name as a historian with his immensely successful history of Scotland and his history of Charles V to become a leading exponent of the Scottish stadial theory of social development, in which society progressed through stages, each more advanced than the last, and each defined by its mode of subsistence, to culminate with the ascendancy of commercial society.[15] He articulated the assumptions of stadial theory most clearly in his three-volume history of America, published in 1777, which examined Spanish colonization of the Americas.[16] He never finished his account of British colonization, leaving behind a fragment on the history of New England and Virginia that only reached up to 1688,

---

[13] On Chalmers's belief in the importance of tradition and custom, see Schutz, "George Chalmers," 55, 57. On Chalmers's belief in the importance of commerce, see Cockcroft, *The Public Life of George Chalmers*, 27, 91.

[14] George Chalmers, *Political Annals of the Present United Colonies, from Their Settlement to the Peace of 1763* (London: J. Bowen, 1780), 1:99; George Chalmers, *Political Annals*, Vol. 2, in *Collections of the New York Historical Society* (Trow & Smith, 1868), 1:162. On Chalmers's mixed view of New England, see Craven, *The Legend of the Founding Fathers*, 52.

[15] For background on Robertson, see Jeffrey R. Smitten, *The Life of William Robertson: Minister, Historian, and Principal* (Edinburgh: Edinburgh University Press, 2017); and Stewart J. Brown, "William Robertson and the Scottish Enlightenment," in *William Robertson and the Expansion of Empire*, ed. Stewart J. Brown (Cambridge: Cambridge University Press, 1997), 7–35. On Scottish stadial theory, see Karen O'Brien, *Narratives of Enlightenment: Cosmopolitan History from Voltaire to Gibbon* (Cambridge: Cambridge University Press, 1997), 132–36; J. G. A. Pocock, *Barbarism and Religion: Narratives of Civil Government* (Cambridge: Cambridge University Press, 1999), 2:309–29; Ronald Meek, *Social Science and the Ignoble Savage* (Cambridge: Cambridge University Press, 1976); and H. M. Hopfl, "From Savage to Scotsman: Conjectural History in the Scottish Enlightenment," *Journal of British Studies* 17 (1978): 19–40.

[16] On how this work articulated the basic assumptions of stadial theory, see László Kontler, *Translations, Histories, Enlightenments: William Robertson in Germany, 1760–1795* (New York: Palgrave Macmillan, 2014), 125–40; and O'Brien, *Narratives of Enlightenment*, 153–60.

published after his death by his son in 1796. Robertson drew from a number of Colonial and loyalist histories, including those by Thomas Hutchinson, Robert Beverly, and William Stith, for this work, but Chalmers's *Annals* was one of his most important and frequently cited sources. Robertson for the most part did not plagiarize from Chalmers in the way that Marshall did, paraphrasing Chalmers more fully into his own words and generally providing citations with page numbers. Like Chalmers, Robertson sought to balance his vision of commercial progress with his desire for social order, but he endeavored to achieve this goal through a commitment to impartiality. Believing that the best way to maintain order and control was by balancing different perspectives, Robertson strove for this balance in his political and religious positions as well as in his historical writing.[17]

Hence, although he agreed with Chalmers in the main lines of his interpretation of Colonial history, emphasizing the spirit of independence and innovation that characterized the New England colonies, he toned down or omitted Chalmers's scathing attacks on the colonists' fractiousness. And where Chalmers threaded the theme of commercial prosperity into his discussion of all the colonies, Robertson put more emphasis on this theme in his treatment of Virginia, which centered on the obstacles and stimulants to the colony's economic growth, bringing it up only occasionally in his discussion of New England. The result was to mute the tension between order and commercial prosperity so evident in Chalmers's account, though his decision to leave his history unfinished suggests that he, too, in the end, was unable to resolve this tension. A defender of British supremacy over the colonies who at the same time expressed sympathy for their rights as Englishmen and believed they would eventually become independent, Robertson abandoned his history in 1781, finding it difficult to maintain the impartiality he considered so necessary for social order in the increasingly contentious atmosphere of the Revolutionary era.[18]

Marshall's reliance on Chalmers and Robertson stemmed in part from his shared desire to balance a dynamic vision of commercial progress with social order, but he reinterpreted them to show how these two imperatives could be reconciled with one another. Marshall revealed how his commit-

---

[17] On Robertson's commitment to impartiality, see Jeffrey Smitten, "Impartiality in Robertson's History of America," *Eighteenth-Century Studies* 19, no. 1 (1985): 56–77; and Jeffrey Smitten, "Moderatism and History: William Robertson's Unfinished History of British America," in *Scotland and America in the Age of the Enlightenment*, eds. Richard B. Sher and Jeffrey Smitten (Edinburgh: Edinburgh University Press, 1990), 166–70.

[18] Smitten, "Moderatism and History," 164–75. See also Neil Hargraves, "Enterprise, Adventure and Industry: The Formation of 'Commercial Character' in William Robertson's History of America," *History of European Ideas* 29, no. 1 (2003): 47–54, on the tensions between commercial order and disorderly adventurism in Robertson's history.

ment to the Scottish Enlightenment vision of commercial progress fueled by individual enterprise and industry contributed to his plagiarism from them in his discussion of Virginia's change from a system of communal property to one based on individual ownership.[19] Citing both Chalmers and Robertson, Marshall echoed them in attributing Virginia's early scarcities to its policy of owning property in common, and like them, viewed the institution of private property as a turning point in putting the colony on a path to prosperity. Drawing on Robertson's account for his analysis of how the acquisition of private property was necessary as an incentive to industry, Marshall explained, "Industry itself, deprived of its due reward, exclusive property in the produce of its toil; felt no sufficient stimulus to exertion."[20] Consequently, Sir Thomas Dale's decision to divide the land into small private plots brought about a dramatic increase in the colonists' productivity, and, "[i]ndustry, having from this moment the certain prospect of recompense, advanced with rapid strides, and the inhabitants were no longer in fear of wanting bread, either for themselves, or for the emigrants who came annually from England."[21] Copying directly from Robertson for some of his language, Marshall here embraced the basic tenets of the Scottish Enlightenment theory of social progress in his emphasis on how the advent of private property invigorated the enterprise and industry necessary for commercial growth by providing individuals with an incentive to exertion.[22]

Marshall likewise revealed his commitment to these tenets in his appropriations from Chalmers. Like Chalmers, Marshall emphasized New England's economic success, drawing from him when he attributed their "unexampled state of prosperity" to the "perfectly free and unrestrained commerce" they enjoyed under Cromwell.[23] Marshall also echoed Chalmers when he described the "vigour and sagacity" Massachusetts showed in its external affairs.[24] But Chalmers immediately went on to provide a lengthy

---

[19] On Marshall's faith in commerce and private enterprise, see Gilbert, "John Marshall," 192–94. See Peter Messer, *Stories of Independence: Identity, Ideology, and Independence in Eighteenth-Century America* (DeKalb: Northern Illinois University Press, 2005), 7–11, 19–27, 73–77; and Arthur Shaffer, *The Politics of History: Writing the History of the American Revolution, 1783–1815* (Chicago: Precedent Publishing, 1975), 77–78, on the influence of Scottish Enlightenment thought on American historical writing in this period.

[20] Marshall, *The Life of George Washington*, 1:52; William Robertson, *The History of America, Books IX and X: Containing the History of Virginia to the Year 1688, and New England to the Year 1652* (London: A. Strahan, T. Cadell Jr., and W. Davies, 1796), 96.

[21] Marshall, *The Life of George Washington*, 1:52; Robertson, *The History of America*, 97.

[22] On the role of industry and enterprise in Robertson's vision of commercial society, see Hargraves, "Enterprise, Adventure and Industry," 39–42, 49.

[23] Chalmers, *Political Annals*, 1:192; Marshall, *The Life of George Washington*, 1:149.

[24] Chalmers, *Political Annals*, 1:188.

and cutting analysis of the colonists' "abominable persecutions" of dissenters, showing how that vigor expressed itself in a corrosive fanaticism and
pointing to the thin line that separated these traits.[25] In contrast, Marshall
followed his reference to the colonists' vigor and sagacity by describing
the "industry and attention to their interests" through which they "improved
the advantages afforded them by the temper of the times." The resulting
prosperity enabled the colonists to develop "a degree of strength and
consistence" that enabled these "sober industrious people" to withstand
the later difficulties they faced.[26] In highlighting the colonists' consistency
and sobriety, Marshall showed how the freedom and vigor that made their
economic prosperity possible was tempered by reason and self-control,
directly contrary to Chalmers's emphasis on their ungovernable fanaticism
and intolerance. He thereby made them embody the compatibility between
commercial growth and social order, rather than the tensions between
them.[27]

    And where Marshall emphasized the colonists' own efforts and virtues
as the source of their prosperity, Chalmers ultimately attributed that prosperity to external factors beyond their control—"plenty of good land, to
be obtained easily by all; and freedom to manage their own affairs in the
manner most agreeable to themselves." In pointing to how these causes
"have at all times proved extremely favorable to the growth of such establish-
ments," Chalmers explained New England's prosperity in terms of the
universal principles that characterized Enlightenment philosophical history.[28] Sharing in the goal of philosophical history to lay the basis for a
science of society through a systematic inquiry into the general laws and
principles that governed human nature, he repeatedly interpreted the events
he described in terms of these principles and thereby turned his analysis
of the colonies into an illustration of broader precepts.[29] By omitting this

---

[25] Chalmers, *Political Annals*, 1:191. See Messer, *Stories of Independence*, 56–58, on Chalmers's critique
of the destabilizing effects of the New England Puritans' religious zeal. On the thin line between constructive
vigor and destructive zeal, see Philip Gould, *Covenant and Republic: Historical Romance and the Politics
of Puritanism* (Cambridge: Cambridge University Press, 1996), 47–52, 172–209.

[26] Marshall, *The Life of George Washington*, 1:149.

[27] On how Marshall believed it was possible to channel and check such vigor in his legal theory, see R.
Kent Newmyer, *John Marshall and the Heroic Age of the Supreme Court* (Baton Rouge: Louisiana State
University Press, 2001), 264–65.

[28] Chalmers, *Political Annals*, 1:192.

[29] On philosophical history, see Neil Hargraves, "The 'Progress of Ambition': Character, Narrative, and
Philosophy in the Works of William Robertson," *Journal of the History of Ideas* 63 (April 2002): 262–63;
Mark Salber Phillips, *Society and Sentiment: Genres of Historical Writing in Britain, 1740–1820* (Princeton,
NJ: Princeton University Press, 2000), 48–59; Karen O'Brien, *Narratives of Enlightenment: Cosmopolitan
History from Voltaire to Gibbon* (Cambridge: Cambridge University Press, 1997), 9; and Pocock, *Barbarism*,
2: 4–25.

passage, and others like it throughout his work, Marshall divested his account of the attributes of philosophical history and suggested that the colonists' success was the product of their own abilities, rather than of general laws. Marshall thus made the colonists' achievements appear to be specific to them, fueling the belief in American uniqueness that underpinned American exceptionalism.[30]

Marshall's rejection of philosophical history furthered his nationalist purposes in another way. In stripping away Chalmers's appeals to philosophical history, Marshall also vindicated the colonists from Chalmers's attacks on them, for Chalmers based those attacks on the universal principles he derived from his perspective as a philosophical historian. Chalmers interpreted Bacon's Rebellion from that perspective and expressed its assumption of the uniformity of human nature in the parallels he made between the rebellion and the Revolution. Concurring with the "wise" that "the pretenses and practices of insurrection are at all times and in every country the same," Chalmers highlighted the similarities between Bacon's manifesto justifying the rebellion and the proclamations of the revolutionaries when he spoke of how "[l]ike recent declarations, that conduct of the governor, which was the necessary result of their own misconduct, was insisted on as the principal cause of their revolt." "And, as the practices of men placed in similar situations are always the same," the rebels won many English supporters "because the insurrection added to the vexation and embarrassments of their sovereign." [31] Here, Chalmers made broader generalizations about the nature of rebellions to condemn both Bacon's Rebellion and the Revolution as groundless, showing how in both cases, the insurgents provoked the very grievances they protested through their own conduct, while their English supporters used these uprisings to undermine the king.

Although Marshall closely followed both the content and language of Chalmers's description of Bacon's manifesto and agreed with him in abhorring the destructiveness of the rebellion, he left out Chalmers's larger conclusions about its similarities to the Revolution. He thereby implicitly disputed Chalmers's portrayal of the Revolution as a baseless and destructive insurrection akin to Bacon's Rebellion, enabling him to reconcile his hostility to social disorder with his desire to vindicate the Revolution by differentiating it as an orderly and principled struggle against arbitrary

---

[30] For Marshall's aversion to philosophical generalizations, see Shaffer, *The Politics of History*, 156. On the opposition between an exceptionalist narrative of American history and the universalizing perspective of philosophical history, see O'Brien, *Narratives of Enlightenment*, 204–33.

[31] Chalmers, *Political Annals*, 1:334.

authority.[32] At a point when the heated party conflicts of the 1790s and the ascendancy of the Democratic Republicans in the election of 1800 had made him increasingly uneasy about the threat of popular upheaval in his own time, it was all the more important for Marshall, a staunch Federalist, to ward off that danger by disassociating the Revolution from uprisings like Bacon's Rebellion and showing how it was grounded in respect for law.[33]

Hence he disputed Chalmers's indictment of the Revolution as the culmination of a long pattern of illegal claims to authority by the colonists. He did so, however, simply by softening or by leaving out Chalmers's denunciations of the colonists' fractiousness without offering any judgments or commentary of his own, thereby making his defense of them appear to be the impartial expression of truth rather than his own opinion.[34] Yet in tempering the rancor of Chalmers's analysis to give his account a more impartial tone, Marshall was himself following Robertson's lead. Robertson's work thus provided him with a template for how to reinterpret Chalmers and converted Chalmers's account into a form that lent itself more easily to his nationalist purposes. For example, Marshall minimized the Massachusetts colonists' infringements of both English law and the rights of dissenters by relying on Robertson's discussion of the independent church they created, which itself paraphrased from Chalmers's account while softening his criticisms of them. Citing Robertson as his sole source, Marshall echoed him in his description of how the Massachusetts Puritans "concurred in the institution of a church, in which was established that form of policy, which was believed best to agree with the divine will as revealed in the scriptures, and which has since been denominated independent."[35] What Marshall omitted from Robertson's analysis was his reference to how the Massachusetts colonists established this church without any heed to the king, upon whose authority the colony was based, and "in contempt of the laws of England, with which the charter required that none of their acts or ordinances should be inconsistent."[36] By leaving out Robertson's comment about how their creation of an independent church violated their charter and English law, Marshall made the Massachusetts colonists appear less insubordinate than they did in Robertson's account.

But in making this statement, Robertson had himself condensed and softened Chalmers's scathing analysis of what a total claim to sovereignty

---

[32] Marshall, *The Life of George Washington*, 1:190–91. On Marshall's hostility to social disorder and his proclivity for the orderly side of the Revolution, see Gilbert, 187. On how the revolutionary historians shared this concern, see Shaffer, *The Politics of History*, 123–26, 140–41.

[33] On Marshall's unease about democratic politics, see Messer, *Stories of Independence*, 163–65.

[34] On this strategy, see Cheng, "Plagiarism in Pursuit of Historical Truth," 151–52.

[35] Marshall, *The Life of George Washington*, 1:96.

[36] Robertson, *The History of America*, 199.

the formation of this church represented. Chalmers went beyond pointing to the illegality of the colonists' actions to suggest that they were set on independence from the start when he remarked on how impossible it was "to support the legality of the association before-mentioned; except on principles of pure independence."[37] And where Robertson simply noted the colonists' violation of English law without giving his own judgment of their actions, Chalmers made clear his disapproval of them as he mocked the self-righteousness that enabled them to disregard the system of English law that he applauded, observing caustically, "The laws of England, so justly celebrated by the panegyric of nations, they considered as not binding on them because inapplicable to so godly a people."[38] In moderating and shortening this condemnation, Robertson provided a version of these events more acceptable to Marshall and made it easier for Marshall to leave out any reference to the illegality of the colonists' actions at all.[39]

In his discussion of the requirements for church membership and communion, Marshall likewise cited and drew from Robertson, who in turn cited and copied from Chalmers for his analysis.[40] According to Chalmers, "And none was admitted to communion with them without giving satisfaction to the church concerning his faith and manners. But the mode how that should be given was left to the arbitrary discretion of the elders, as particular cases should arise: Thus erecting in wilds, which freedom was to people and cultivate, that inquisitorial power which had laid waste the fruitfullest European plains."[41] Robertson, however, omitted the last part of this passage criticizing the elders' authority to determine who was eligible to receive communion as an "inquisitorial power" that went counter to the spirit of freedom that inspired colonization, and Marshall followed suit. Marshall thus turned to Robertson to exonerate the colonists from Chalmers's portrayal of them as tyrannical persecutors and used him to downplay the illegality of the colonists' actions, thereby implicitly affirming their commitment to liberty and showing how that liberty was compatible with respect for the law.

In other places, Robertson departed altogether from Chalmers's account, providing Marshall with an alternative perspective that he used as a counter to Chalmers's analysis. For example, Robertson put more empha-

---

[37] Chalmers, *Political Annals*, 1:144.

[38] Chalmers, *Political Annals*, 1:144.

[39] On how Robertson tempered Chalmers's more critical view of the Puritan colonists, see Smitten, "Moderatism and History," 169–70.

[40] Marshall, *The Life of George Washington*, 1:97; Robertson, *The History of America*, 200; Chalmers, *Political Annals*, 1:143.

[41] Chalmers, *Political Annals*, 1:143.

sis on Native American savagery and brutality against the colonists than Chalmers did, enabling Marshall to use his work as an instrument to challenge Chalmers's more critical portrayal of the colonists. While denigrating Native Americans for their "cunning and cruelty" in their "most barbarous massacre" of the Virginia colonists in 1622, Chalmers also recognized the colonists' brutality against Native Americans and their culpability for the violence between them, describing the "consuming war" that "ensued" in which "a repetition of mutual wrongs hath transmitted to their posterity mutual abhorrence." Hence, "[w]hen the Indians of the present day would express their hatred and fear of the Virginians, they call them 'the long knife.'" Chalmers further acknowledged the colonists' culpability for the conflict by noting their failure to conciliate the Indians and their dispossession of the Indians from their lands without their consent or any compensation, admitting that they "had never been solicitous to cultivate the good-will of the aborigines; and had neither asked permission when their country was occupied, nor had given a price for invaluable property, which was taken without authority."[42]

Marshall refuted this criticism of the colonists' disregard for the Indians and justified their treatment of them by turning to Robertson's account of the war, which drew on William Stith's history of Virginia to portray the Indians more unfavorably than Chalmers did. Marshall followed Robertson closely as he described the close and peaceful interactions between the colonists and the Indians that preceded the massacre and emphasized how unsuspecting and trusting the colonists were in allowing such intermingling. Marshall copied much of Robertson's language as he wrote of how "the Indians, being often employed as hunters, were furnished with fire arms, and taught to use them. They were admitted at all times freely into the habitations of the English, as harmless visitants." Thus drawing from Robertson to paint the colonists as innocent and trusting victims of vengeful and calculating Indians, Marshall paraphrased him to bring this contrast into even stronger relief as he wrote of how it was "[d]uring this state of free and friendly intercourse" that they "formed, with cold and unrelenting deliberation, the plan of a general massacre, which should involve man, woman, and child in indiscriminate slaughter."[43]

---

[42] Chalmers, *Political Annals*, 1:58–59. On Chalmers's sympathy for Native Americans and his criticism of the colonists' treatment of them, see Messer, *Stories of Independence*, 64.

[43] Marshall, *The Life of George Washington*, 1:64; Robertson, *The History of America*, 106. On Robertson's negative portrayal of Native Americans as cruel savages, see Stewart Brown, "An 18th-Century Historian on the Amerindians: Culture, Colonialism and Christianity in William Robertson's History of America," *Studies in World Christianity* 2 (1996): 212–13. On Marshall's plagiarism for this passage, see Foran, "John Marshall," 55–56.

Marshall omitted, however, Robertson's lengthy condemnation of the colonists' brutality against the Indians in retaliation for this attack, which balanced out his depiction of Native American ruthlessness.[44] This brutality was so great that Robertson compared it to Spanish atrocities against the Indians, as the English colonists became "regardless like them of those principles of faith, honour, and humanity" that restrained "civilized nations" in their desire for vengeance, to the point where they became as savage as the Indians. Consequently, "the behaviour of the two people seemed now to be perfectly reversed," as "the English, with perfidious craft, were preparing to imitate savages in their revenge and cruelty."[45] Marshall, in contrast, downplayed and justified the colonists' cruelty to the Indians by only briefly mentioning it and emphasizing that it was in retaliation for the atrocities that the Indians committed against them. Thus Marshall spoke of the "vindictive and exterminating war" that followed "this horrible massacre," "in which, were successfully practised upon the Indians the wiles, of which they had set so bloody an example. During this disastrous period, many public works were abandoned; the college institution was deserted; the settlements were reduced from eighty to eight; and famine superadded to the accumulated distresses of the colony, its afflicting scourge."[46] And by pointing to the sufferings and destruction that the colonists endured, Marshall again painted them as victims of Indian aggression rather than active agents of brutality against the Indians. He thereby justified not only the colonists' reprisals against them, but also the process of conquest and expansion that had created the nation and that was becoming an increasingly important part of America's exceptionalist mission in his own time.[47]

But even as his emphasis on Native American brutality sanctioned their dispossession and exclusion from America's destiny to expand westward, he at the same time shied away from defining that exclusion in racial terms and making it categorical.[48] While Marshall painted the colonists

---

[44] On Robertson's effort to provide an impartial assessment of Native Americans and their European conquerors, see Smitten, "Impartiality in Robertson's *History of America*," 58–65.

[45] Robertson, *The History of America*, 111–12.

[46] Marshall, *The Life of George Washington*, 1:65–66.

[47] For Marshall's views on Native Americans, see Newmyer, *John Marshall*, 440–58. On the importance of expansion to American identity in this period and its implications for the treatment of Native Americans, see Robert Berkhofer, *The White Man's Indian: Images of the American Indian from Columbus to the Present* (New York: Vintage Books, 1978), 134–66. On the fascination with Native American violence, see Bernard W. Sheehan, *Seeds of Extinction: Jeffersonian Philanthropy and the American Indian* (Chapel Hill: University of North Carolina Press, 1973), 185–212.

[48] On how American expansionist rhetoric had not yet become racialized at this point, see Reginald Horsman, *Race and Manifest Destiny: The Origins of American Racial Anglo-Saxonism* (Cambridge, MA: Harvard University Press, 1981), 1–6, 98–115.

as innocent victims of Native American brutality even more than Robertson did, he omitted Robertson's references to how such brutality was the product of their savage nature, stripping his appropriations from Robertson of their philosophical component. Where Robertson described how the Indians "murdered men, women, and children, with undistinguishing rage, and that rancorous cruelty with which savages treat their enemies," Marshall simply wrote that they "murdered without distinction of age or sex," leaving out Robertson's reference to the "rancorous cruelty" that characterized "savages."[49] In making these generalizations about "savage" cruelty, Robertson employed the language of stadial theory, which used the category of "savage" to describe the earlier, more "primitive" stage of social development. By explaining Native American behavior in terms of the general categories of stadial theory, Robertson placed his analysis of the massacre within the broader framework of philosophical history. What Robertson's language left ambiguous here was whether savagery—and the brutality he associated with it—was inherent to Native Americans, or whether they could advance beyond that state to another stage of society.[50] In divesting his account of Robertson's broader claims about savagery, then, Marshall also divested it of their potentially racialist implications; the brutality he described was historically specific, not something inherent to Native Americans in general, leaving open the possibility of assimilating them into White American culture once they were no longer a threat.[51]

Marshall likewise repudiated Robertson's defense of slavery's benefits to offer a more equivocal assessment of its impact on Virginia. Robertson portrayed slavery as an engine of prosperity for Virginia as he highlighted the arrival of women and of slaves as "two events" through which the

---

[49] Robertson, *The History of America*, 109; Marshall, *The Life of George Washington*, 1:65.

[50] On Robertson's application of stadial theory to Native Americans, see O'Brien, *Narratives of Enlightenment*, 156–61; Kontler, *Translations*, 126–40; and Bruce Lenman, "'From savage to Scot' via the French and the Spaniards: Principal Robertson's Spanish Sources," in *William Robertson and the Expansion of Empire*, ed. Stewart Brown (Cambridge: Cambridge University Press, 1997), 196–209. On Robertson's belief in the humanity of Native Americans and his doubts about their capacity to advance to a more civilized state, see Brown, "An Eighteenth-Century Historian on the Amerindians," 218–20. On how Robertson generalized about Native Americans as a "race," see Nicholas Hudson, "From 'Nation' to 'Race': The Origin of Racial Classification in Eighteenth-Century Thought," *Eighteenth-Century Studies* 29 (Spring 1996): 250. For a different view that emphasizes Robertson's environmentalism, see Roy Harvey Pearce, *Savagism and Civilization: A Study of the Indian and the American Mind*, rev. ed. (1953; Berkeley: University of California Press, 1988), 86–88. For the dualities in Robertson's portrayal of Native Americans, see Pocock, *Barbarism and Religion*, 4: 195–204.

[51] On the racialist implications of Scottish stadial theory, see Roxann Wheeler, *The Complexion of Race: Categories of Difference in Eighteenth-Century British Culture* (Philadelphia: University of Pennsylvania Press, 2000), 181–90. On Marshall's belief in the possibility of assimilation and his sympathy for the current plight of Native Americans, see Newmyer, *John Marshall*, 441–43; and Joel Richard Paul, *Without Precedent: Chief Justice John Marshall and His Times* (New York: Riverhead Books, 2018), 401–04.

colony's "population and industry were greatly promoted." By providing the colonists "with means of executing" "with greater facility" the "more extensive plans of industry" that came about as the presence of women stabilized the male colonists, slavery was an important adjunct in furthering the commercial progress launched by the arrival of women.[52] Hence, despite his own belief in its wrongfulness, Robertson emphasized the value of slavery to the colony, for, "as that hardy race was found more capable of enduring fatigue under a sultry climate than Europeans," "their aid seems now to be essential to the existence of the colony, and the greater part of field labour in Virginia is performed by servile hands."[53] Here, Robertson generalized about the differences between Africans and Europeans as he did with Native Americans, attributing their enslavement to their capacity for labor in warmer climates—a physical difference that could serve as a basis for categorizing Africans as racially distinct.

Marshall disavowed Robertson's potentially racialist vision of American prosperity predicated on the indefinite enslavement of people of African descent by drawing on Chalmers to give a less favorable assessment of slavery's impact on Virginia. Although he cited both Robertson and Chalmers for this section, he adhered more closely to Chalmers in the structure and content of his analysis, as he wrote of how the advent of free trade was "undoubtedly of essential benefit" to Virginia while also pointing to how this development was "the immediate cause of introducing a species of population, which has had vast influence on their past situation, and may affect their future destinies in a manner which human wisdom can neither foresee nor control," with the arrival of a Dutch ship bringing the colony's first importation of African slaves.[54] Where, then, Robertson viewed slavery as an instrument of commercial progress, Marshall followed Chalmers in suggesting that it was a drawback to that progress. But unlike Chalmers, who was explicit in deploring "the sad epoch of the introduction of African slaves into the colonies" as a drawback to the advent of free trade, Marshall only implied that this was the case as he spoke more ambiguously of the "vast influence" of slavery on Virginia's past and its unforeseeable effects on its future.[55] Writing in the wake of Gabriel's Rebellion, which brought home to White Virginians the dangers of slavery and led them to briefly reconsider their commitment to the institution,

---

[52] Robertson, *The History of America*, 99–100.

[53] Robertson, *The History of America*, 101. For Robertson's treatment of slavery, see Hargraves, "Enterprise, Adventure and Industry," 50–51; and Brown, "An Eighteenth-Century Historian on the Amerindians," 210–11.

[54] Marshall, *The Life of George Washington*, 1:61.

[55] Chalmers, *Political Annals*, 1:49.

Marshall here expressed his unease about those dangers but stopped short of condemning slavery outright.[56] Marshall's position as a slaveowner himself made it difficult for him to be as forthright as Chalmers was in his criticism of slavery, despite his view of it as an "evil."[57] Thus he also omitted Chalmers's strictures on the hypocrisy of the colonists, "who had just emerged from a state of slavery themselves," as they "first reduced their fellow-men to the condition of brutes." Loath as he was to follow Robertson in relegating Blacks to a state of indefinite enslavement, then, neither would he go as far as Chalmers did in acknowledging the colonists' culpability for slavery, revealing the limits on both his opposition to the institution and his willingness to include Blacks in his vision of the nation—hence his support for colonization.[58]

In stripping his appropriations from Chalmers and Robertson of their philosophical element, Marshall articulated a vision of social progress that was both more and less exclusionary than theirs. On the one hand, his analysis promoted a chauvinistic nationalism in suggesting that Americans were uniquely able to balance commercial prosperity with social order. On the other hand, even as he excluded Native Americans and African Americans from this vision, he eschewed the generalizations that would have given a racial and permanent basis to that exclusion. Thus if in his plagiarism from Chalmers and Robertson, Marshall provided a medium for disseminating their interpretations to his fellow Americans, it was a considerably revamped version that he gave them. Paradoxically, even as his reliance on Chalmers and Robertson revealed America's continued dependence on its British and loyalist roots, he broke free of those roots by turning the words of his loyalist and British opponents against themselves. Plagiarism in this way served as an important vehicle for the transatlantic exchange of knowledge, enabling Marshall, Robertson, and Chalmers to engage in a three-way conversation with one another, while at the same time acting as a tool in the creation of a distinctively American identity.[59]

---

[56] On the reaction to Gabriel's Rebellion, see Douglas R. Egerton, *Gabriel's Rebellion: The Virginia Slave Conspiracies of 1800 and 1802* (Chapel Hill: University of North Carolina Press, 1993), 132–62.

[57] Paul, *Without Precedent*, 46.

[58] Chalmers, *Political Annals*, 1:49. For Marshall's views on slavery, see Newmyer, *John Marshall*, 414–23; Paul, *Without Precedent*, 45–53; and Jean Edward Smith, *John Marshall: Definer of a Nation* (New York: H. Holt & Co., 1996), 488–90. See also Messer, *Stories of Independence*, 129–30, for Marshall's defense of slavery in other parts of his history.

[59] On the importance of copying to American nationhood, see William Huntting Howell, *Against Self-Reliance: The Arts of Dependence in the Early United States* (Philadelphia: University of Pennsylvania Press, 2015).

# 2

# Reproducing Rotifers: "Working Images" and the Making of a Microscopy Community in the Nineteenth Century

## Lea Beiermann

In 1855, the English naturalist Philip Henry Gosse (1810–88) set out to write a book with the tentative title *The Pond-Raker*, which was meant to provide a "popular introduction to the Rotifera."[1,2] Research into rotifers, a phylum of microscopic animals, had gained traction with naturalists by the mid-nineteenth century, following the publication of Christian Gottfried Ehrenberg's seminal work *Die Infusionsthierchen* (1838) and Andrew Pritchard's *A History of Infusoria* (1842).[3] Moreover, Gosse's work on his manuscript of *The Pond-Raker* coincided with a general trend of microscopy becoming more popular in Britain from the 1850s onward, as microscopy publications began to multiply and microscopes became more affordable.[4] Still, as Gosse's son, Edmund, recalled in 1890, "it proved difficult to popularize so abstruse a subject [as rotifers], and *The Pond-Raker* ... soon quitted his pond and dropped his rake."[5] Gosse never finished his book.

However, in the early 1860s, Gosse started publishing illustrated articles on rotifers in the *Popular Science Review* and over the following decades, reproductions of Gosse's rotifer illustrations were circulated widely. They appeared in children's books; featured in his and Charles Thomas Hudson's major specialist work on rotifers, *The Rotifera* (1886/1889);[6] and were transformed into visual teaching aids by Hudson.[7] Hudson's rotifer illustrations were distributed by Thomas Bolton, whose Birmingham *Microscopist's and Naturalist's Studio* sent microscopic specimens, notes, and illustrations to microscopists in Britain and abroad.[8]

[1] This chapter uses data generated via the Zooniverse.org platform, development of which is funded by generous support, including a Global Impact Award from Google, and by a grant from the Alfred P. Sloan Foundation. The author would like to thank the citizen scientists who contributed to *Worlds of Wonder* on Zooniverse, especially Peter Mason, who helped to analyze the data generated by the project. The author's research was funded by the Dutch Research Council (NWO) as part of the Advancing Microscopy project, grant number PGW.18.033/6344.

[2] Edmund Gosse, *The Life of Philip Henry Gosse* (London: Kegan Paul, Trench, Trübner & Co., 1890), 256.

[3] *Infusoria* was a collective term for aquatic microorganisms, including rotifers.

[4] See William H. Brock, "Patronage and Publishing: Journals of Microscopy 1839–1989," *Journal of Microscopy* 155, no. 3 (1989): 249–66; Gerard L'E. Turner, *The Great Age of the Microscope: The Collection of the Royal Microscopical Society through 150 Years* (London: Taylor & Francis, 1989); Jutta Schickore, *The Microscope and the Eye: A History of Reflections, 1740–1870* (Chicago: University of Chicago Press, 2007).

[5] Edmund Gosse, *Life of Philip Henry Gosse*, 256.

[6] *The Rotifera* was published in six parts between January and October 1886. A supplement was published in 1889.

[7] For a list of Hudson's transparencies, held by the University of Exeter, see Robin Wootton, "The Hudson Transparencies. A Set of Remarkable Visual Aids by a Distinguished Victorian Microscopist," *Report and Ttransactions of the Devonshire Association for the Advancement of Science, Literature and Art* 143 (2011): 61–90.

[8] Thomas Bolton, *Hints on the Preservation of Living Objects and Their Examination under the Microscope* (Birmingham: Herald Printing Offices, 1879–82); "Science-Gossip," *The American Monthly Microscopical Journal* 19 (1898): 14–16.

The rise of microbiological studies in the second half of the nineteenth century has been attributed to factors ranging from the introduction of cheaper scientific instruments and printing techniques, to rising literacy rates and the sheer availability of microorganisms.[9] The fascination for rotifers in particular was spurred by the spread of freshwater aquariums, which could be stocked with rotifers obtained from ponds and puddles.[10] The emergence of freshwater field stations from the late nineteenth century onward finally helped to turn the study of freshwater flora and fauna into a research domain of its own, limnology.[11] This chapter traces the rise of rotifer studies by reconstructing the spread of rotifer illustrations among microscopists, showing how the reproduction and circulation of illustrations shaped the nineteenth-century microscopy community.

Microscopy books and periodicals were often richly illustrated, trying to recreate the "world of wonder and beauty" as seen through the microscope.[12] Lightman has called on historians of nineteenth-century science to take a closer look at the abundant visual culture of Victorian microscopy.[13] To date, however, the literature on the history of nineteenth-century microscopy mainly relies on textual sources, notable exceptions being Belknap's analysis of the collaborative expertise that went into producing illustrations for the *Journal of the Quekett Microscopical Club* and Lightman's study of the microscopy illustrations of John George Wood, a nineteenth-century natural theologian.[14] Neither Belknap nor Lightman have considered the possibility of microscopy images moving between media and the effect this may have had on the microscopy community. By following circulating microscopy images through different media, this chapter shows that reproducing illustrations allowed different scientific writers to appropriate these illustrations and write—or draw—themselves into the microscopy community. This is elaborated in the following method-

---

[9] Charles G. Hussey estimates that in British publications references to rotifers increased more than ten-fold between 1860 and 1890, "An Historical Survey of the Collection and Study of Rotifers in Britain," *Hydrobiologia* 73, no. 1–3 (1980): 237–40.

[10] Bernd Brunner, *The Ocean at Home: An Illustrated History of the Aquarium* (New York: Princeton Architectural Press, 2005).

[11] De Bont argues that the establishment of the freshwater field station in Plön, Germany, in 1891 was crucial in institutionalizing limnology. Raf De Bont, *Stations in the Field: A History of Place-Based Animal Research, 1870–1930* (Chicago: University of Chicago Press, 2015).

[12] Philip Henry Gosse, *Evenings at the Microscope; or Researches among the Minuter Organs and Forms of Animal Life* (New York: D. Appleton, 1860), v; Bernard Lightman, "The Microscopic World," *Victorian Review* 36, no. 2 (2010): 46–49.

[13] Lightman, "The Microscopic World."

[14] Bernard Lightman, "The Visual Theology of Victorian Popularizers of Science: From Reverent Eye to Chemical Retina," *Isis* 91, no. 4 (2000): 651–80; Geoffrey Belknap, "Illustrating Natural History: Images, Periodicals, and the Making of Nineteenth-Century Scientific Communities," *The British Journal for the History of Science* 51, no. 3 (2018): 395–422.

ological section. The remainder of the chapter traces reproductions of rotifer illustrations that preceded or followed from the publication of Gosse and Hudson's influential *The Rotifera* and argues for the importance of these reproductions in the making of the nineteenth-century microscopy community.

## A COOPERATION OF OBSERVERS

Daston and Lunbeck have convincingly argued that "sustained observation creates communities."[15] From the early modern period onward, observers of the natural world were recruited and coordinated over vast distances, which affected both their observations and beliefs of who should be considered a reliable observer. Many nineteenth-century observers of microscopic objects indeed considered themselves part of a community that cut across national, disciplinary and class boundaries. In his annual address of 1866, James Glaisher, then president of the Microscopical Society of London, declared that to advance microscopy a "co-operation of observers scattered all over the world is necessary, and these should include all classes, for so universal are the objects scattered which we wish to study, that a large co-operation is indispensable."[16,17] Achieving Glaisher's envisioned universal cooperation among microscopists was no mean feat, with microscopists being at the same time botanists, geologists, zoologists, mathematicians, engineers, teachers or physicians, connected only by their use of the microscope.[18] Glaisher himself claimed that circulating microscopy publications, crossing disciplinary and national boundaries and inviting their readers to contribute observations, were a useful means of facilitating cooperation among microscopists.[19] Most of these publications, including the journal of Glaisher's Microscopical Society of London, contained drawings of microscopic specimens and instruments, made by microscopists in collaboration with illustrators, engravers, and printers.

In his analysis of nineteenth-century sketches of celestial nebulae, Nasim puts forward the concept of "working images," preliminary scientific

---

[15] Lorraine Daston and Elizabeth Lunbeck, eds., *Histories of Scientific Observation* (Chicago: University of Chicago Press, 2011), 369.

[16] James Glaisher, "The President's Address for the Year 1866," *Transactions of the Microscopical Society of London* 14. New series (1866): 48.

[17] The Microscopical Society of London became the Royal Microscopical Society after receiving its Royal Charter in 1866.

[18] See Matthews's presidential address to the London Quekett Microscopical Club: John Matthews, "President's Address," *The Journal of the Quekett Microscopical Club* 4 (1876): 186–94.

[19] James Glaisher, "The President's Address," *The Monthly Microscopical Journal* 11 (1869): 141–54.

drawings that precede the published illustration and that are themselves a mode of observation, or a way of probing what is seen with the eye.[20] Nasim defines *working images* as messy, private sketches that are part of the practice of scientific observation and that only later result in stabilized published illustrations, or "immutable mobiles."[21] Although this chapter embraces Nasim's proposal of acknowledging the making of illustrations as an observational practice, it undermines his distinction between volatile sketches and allegedly immutable, published illustrations. Because the nineteenth-century print trade thrived on reproducing texts and illustrations, it is virtually impossible to determine what counts as the published end result of a sketch.[22] Even after its first publication, a drawing would often be copied, adapted, and republished, making these illustrations hardly less mutable than the initial drawing.[23] Moreover, some of the sources that the following analysis draws on, for example, a series of flyers that was only later bound and formally published, moved in a liminal space between the published and unpublished.

Recognizing the instability and longer trajectory of published and unpublished scientific illustrations in the nineteenth century broadens the scope of practitioners and materials involved in the making of these illustrations.[24] A scientific illustration was produced not only by the people and objects involved in its first publication but also by those adapting it afterward, copying and republishing it, or by adding the illustration to a collection of journal clippings or a scrapbook.[25] If we take up Nasim's

---

[20] Nasim's "working images" build on Alpers's notion of "picturing" as opposed to "pictures," the allegedly finished, stabilized products of picturing an image. Omar W. Nasim, *Observing by Hand. Sketching the Nebulae in the Nineteenth Century* (Chicago: The University of Chicago Press, 2013). Svetlana Alpers, *The Art of Describing: Dutch Art in the Seventeenth Century* (Chicago: University of Chicago Press, 1983).

[21] Nasim draws on Bruno Latour's famous concept of immutable mobiles in "Drawing Things Together," in *Representation in Scientific Practice*, eds. Michael Lynch and Steve Woolgar (Cambridge, MA: MIT Press, 1990), 19–68.

[22] It was only after the Berne Convention (1886) and the American Chace Act (1891) that copyright laws were enforced internationally.

[23] For an analysis of how Belgian anatomists recycled illustrations, see Veronique Deblon, "Imitating Anatomy: Recycling Anatomical Illustrations in Nineteenth-Century Atlases " in *Bodies Beyond Borders: Moving Anatomies, 1750–1950*, eds. Kaat Wils, Raf de Bont, and Sokhieng Au (Leuven: Leuven University Press, 2017), 115–38.

[24] Latour and Lowe use the term *trajectory* to refer to the life of an artwork. However, they continue to distinguish between an original and its copies, which constitute the "trajectory." As the following analysis shows, this distinction is impossible to maintain in the case of nineteenth-century microscopy illustrations. Bruno Latour and Adam Lowe, "The Migration of the Aura, or How to Explore the Original through Its Facsimiles," in *Switching Codes: Thinking through Digital Technology in the Humanities and the Arts*, eds. Thomas Bartscherer and Roderick Coover (Chicago: University of Chicago Press, 2011), 275–98.

[25] Rose Roberto, "Democratising Knowledge and Visualising Progress: Illustrations from Chambers's Encyclopaedia, 1859–1892" (PhD thesis, University of Reading, 2018). See also Ellen Gruber Garvey, *Writing with Scissors: American Scrapbooks from the Civil War to the Harlem Renaissance* (Oxford: Oxford University Press, 2012).

suggestion of regarding working images as a mode of scientific observation, then the multiple reproductions of microscopy illustrations—by hand and through print—show that scientific observation was, in fact, a drawn-out process that continued even without the microscopic object initially observed. At the same time, as I will show, reproducing and circulating microscopy illustrations made it possible for microscopists to negotiate who should be considered part of the microscopy community.

Over the past decade, concepts of "circulation"—describing the trajectories of material artefacts—have gained currency among global historians of science, technology, and medicine.[26] In a similar vein, researchers in the field of media studies and communication have been putting forward elaborate theories of image circulation.[27] It is primarily the latter strand of literature that informs my analysis. In particular, I build on the method of "iconographic tracking," originally developed by the media studies scholar Laurie Gries to trace viral images online. *Iconographic tracking* refers to following an image through its various material transformations and encounters with people and objects by combining traditional qualitative research with online search tools.[28] Through iconographic tracking, researchers can reconstruct an image's "occasions of use" and can analyze how the image creates assemblages of people and objects.[29] Although one must be careful not to conflate nineteenth-century visual reproductions and present-day viral images, the many digitized nineteenth-century sources that are readily available today lend themselves to being analyzed with digital methods.[30]

---

[26] For example, Stefanie Gänger, *Relics of the Past. The Collecting and Study of Pre-Columbian Antiquities in Peru and Chile, 1837–1911* (Oxford: Oxford University Press, 2014); Kapil Raj, "Beyond Postcolonialism … and Postpositivism: Circulation and the Global History of Science," *Isis* 104, no. 2 (2013): 337–47; Lissa Roberts, "Situating Science in Global History: Local Exchanges and Networks of Circulation," *Itinerario* 33, no. 1 (2009): 9–30; James Poskett, *Materials of the Mind: Phrenology, Race, and the Global History of Science, 1815–1920* (Chicago: University of Chicago Press, 2019).

[27] Some of the most notable works are Cara Finnegan, "Studying Visual Modes of Public Address. Lewis Hine's Progressive-Era Child Labor Rhetoric," in *The Handbook of Rhetoric and Public Address*, eds. Shawn J. Parry-Giles and J. Michael Hogan (Malden, MA: Wiley-Blackwell, 2010), 250–70; Lester C. Olson, *Benjamin Franklin's Vision of American Community: A Study in Rhetorical Iconology* (Columbia: University of South Carolina Press, 2004); Laurie E. Gries, *Still Life with Rhetoric: A New Materialist Approach for Visual Rhetorics* (Logan: Utah State University Press, 2015).

[28] Laurie E. Gries follows a viral image as it "shifts from, among other things, an illustration to propaganda to a genre of critique to a touchstone for copyright law and remix" (338) in her article "Iconographic Tracking: A Digital Research Method for Visual Rhetoric and Circulation Studies," *Computers and Composition* 30, no. 4 (2013): 332–48. For a detailed description of her methodology, see Laurie E. Gries, *Still Life with Rhetoric*.

[29] Gries, *Still Life with Rhetoric*, 338.

[30] For an overview of the challenges of digitizing, and researching digitized, nineteenth-century periodicals and illustrations, see James Mussell, *The Nineteenth-Century Press in the Digital Age* (Basingstoke: Palgrave Macmillan, 2012); Julia Thomas, *Nineteenth-Century Illustration and the Digital*, ed. Andrew Stauffer, Studies in Word and Image (Basingstoke: Palgrave Macmillan, 2017).

The reproductions of rotifer illustrations examined in this chapter were identified through traditional archival research at the Oxford History of Science Museum, full-text searches of the Internet Archive and the Biodiversity Heritage Library, and data generated by two crowdsourcing projects, *Science Gossip* and *Worlds of Wonder*. *Science Gossip*, launched by the ConSciCom research group on Zooniverse in 2015, invited citizen scientists to classify illustrations in nineteenth-century natural history periodicals and identify their contributors.[31] The data generated from three microscopy-related periodicals, *Hardwicke's Science Gossip*, the *Journal of the Royal Microscopical Society*, and the *Journal of the Quekett Microscopical Club* were particularly useful for my analysis.[32] *Worlds of Wonder* invited citizen scientists to help analyze nineteenth-century microscopy publications. Moreover, *Worlds of Wonder* sought to further develop Gries's method of iconographic tracking by asking citizen scientists to flag reproductions of illustrations. It was launched in April 2019 and generated data on the illustrations in several American, British, and German microscopy periodicals; a collection of flyers; and handbooks. Taken together, the data gathered through these traditional and digital archival research methods allowed me to trace the transformations of rotifer illustrations, their various encounters with microscopists, and the consequences for the microscopy community that followed from these encounters.

## CIRCULATING ROTIFER ILLUSTRATIONS

Although Philip Henry Gosse abandoned his plan of writing a popular introduction to rotifers in the mid-1850s, he continued to publish articles on these microscopic animals in various journals and included chapters on rotifers in *Tenby: A Seaside Holiday* (1856) and in his widely read introduction to microscopy, *Evenings at the Microscope* (1859). Gosse illustrated his writings on rotifers himself. With his father an impoverished gentleman trying to make a living off of miniature painting, Gosse had made the drawing of plants and animals a habit when he was only a boy.[33] His popular writings mostly portrayed microscopy as a moral as much as

---

[31] Constructing Scientific Communities—Citizen Science in the 19th and 21st Centuries (ConSciCom, https://conscicom.org/) was an AHRC funded project based at the universities of Leicester and Oxford. Zooniverse is one of the largest citizen science platforms (https://www.zooniverse.org/).

[32] Whereas all nineteenth-century volumes of *Hardwicke's Science Gossip* and the *Journal of the Quekett Microscopical Club* have been analysed by citizen scientists, of the *Journal of the Royal Microscopical Society* only the 1879–1889 (and 1900) volumes have been analysed so far.

[33] Ann Thwaite, *Glimpses of the Wonderful: The Life of Philip Henry Gosse, 1810–1888* (London: Faber & Faber, 2002).

a scientific exercise, allowing the microscopist to marvel at God's most minute creations. However, the journals he published in were aimed at very diverse readerships and Gosse seems to have switched quite effortlessly between different styles of writing and drawing. Whereas *Tenby* presented a lively account of Gosse's rambles along the Welsh shore and the plants and animals he encountered, a paper of his published in the *Philosophical Transactions of the Royal Society of London* in the same year was a densely written treatise on the manducatory organs of the Rotifera.[34] *Tenby* tried to charm its readers with impressive colored plates of rotifers, but it was the simple line drawings in the Royal Society paper and Gosse's *Evenings at the Microscope* that were soon reproduced in other publications.

A search for mentions of Gosse as illustrator in the data gathered through *Worlds of Wonder* reveals that both Henry James Slack's *Marvels of Pond-Life* (1861) and Mary Ward's *Microscope Teachings* (1866) copied one of Gosse's illustrations from the 1856 Royal Society paper and credited him as illustrator (Figure 2.1). *Marvels of Pond-Life* introduced beginners in microscopy to freshwater flora and fauna, each chapter focusing on one month of the year when the reader was most likely to find the specimens described. Most of the illustrations in the book written by the journalist Slack were made by his wife, Charlotte Mary Slack, who combined one of Gosse's illustrations of the jaw of *Floscularia ornata* with illustrations of her own.[35] The resulting plate was later republished in Ward's *Microscope Teachings*, still crediting Gosse for illustrating the *Floscularia* jaw.[36] Ward's *Microscope Teachings*, as well as her other works on microscopy, framed its microscopy illustrations as resulting from observations made in Ward's very own way.

Mary Ward (1827–1869), born into an Anglo-Irish family at Ballylin, was not only geographically at the margins of the British scientific community.[37] As a woman, she did not receive a formal scientific education, but her parents encouraged her in her studies of natural history. After being given a microscope by her father when she was eighteen years old, Ward began to dedicate much of her time to making microscopic observations, as well as describing and illustrating them. She profited from

---

[34] Philip Henry Gosse, "On the Structure, Functions, and Homologies of the Manducatory Organs in the Class Rotifera," *Philosophical Transactions of the Royal Society of London* 146 (1856): 419–52.

[35] Henry James Slack, *Marvels of Pond-Life; or, a Year's Microscopic Recreations among the Polyps, Infusoria, Rotifers, Water-Bears, and Polyzoa* (London: Groombridge and Sons, 1861), viii.

[36] Since *Microscope Teachings* was issued by the same publisher as *Marvels of Pond-Life*, it probably reused the engraving made by Charlotte Mary Slack.

[37] The biographical information on Ward given in this chapter is based on an article by Owen G. Harry, "The Hon. Mrs Ward (1827–1869) Artist, Naturalist, Astronomer and Ireland's First Lady of the Microscope," *The Irish Naturalists' Journal* 21, no. 5 (1984): 193–200.

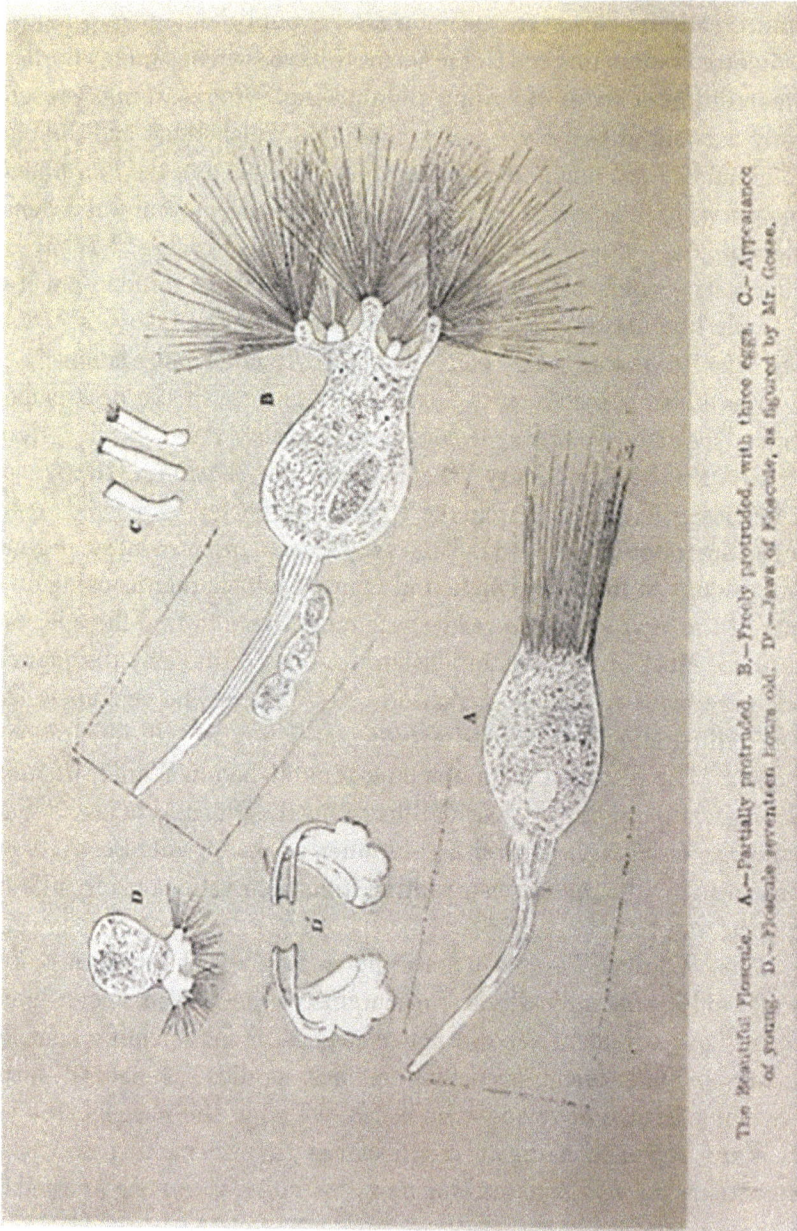

Figure 2.1 Plate in Henry James Slack's *Marvels of Pond-Life*, including an illustration made by Philip Henry Gosse (*D'*).

Image from the Biodiversity Heritage Library. Contributed by the Thomas Fisher Rare Book Library, University of Toronto. *Right:* Plate reproduced in Mary Ward's *Microscope Teachings*. Image from HathiTrust. Contributed by University of Illinois at Urbana—Champaign.

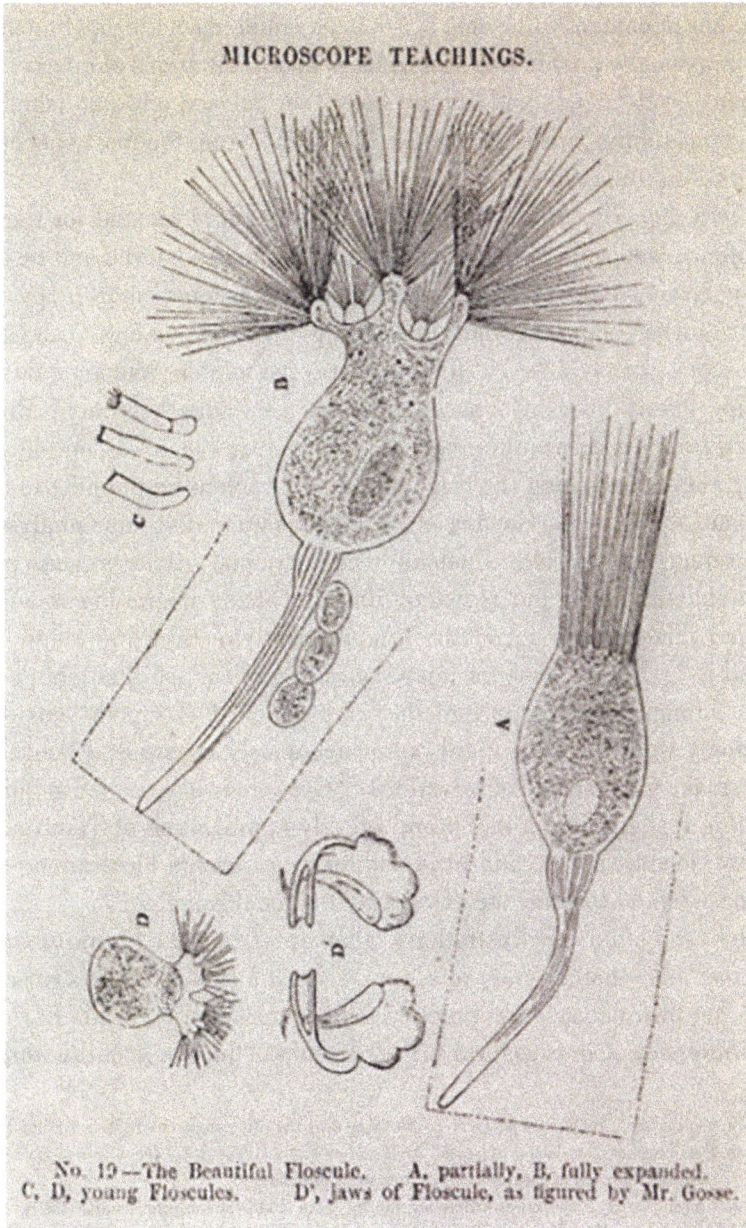

Figure 2.1 (*continued*). Same image reproduced in Mary Ward's *Microscope Teachings*.

the scientific network of her cousin, the astronomer William Parsons, who brought her in contact with some of the most eminent scientists of the time. Ward occasionally passed her descriptions and illustrations of microscopic specimens on to friends and family. Later, she revised and published her microscopic studies, with *Microscope Teachings* being the most successful among these publications.[38]

*Microscope Teachings* combined an introductory manual for the use of the microscope with descriptions of the specimens observed and instructions on how to procure them. Ward's book presented microscopy as a domestic, almost maternal affair, instructing its readers on how to construct a microscope objective for children and observe microscopic animals kept in a wine glass.[39] Gooday has argued that the books written by British popularizers of microscopy aimed at disciplining the lower middle and working classes, advising them to confine their scientific pursuits to their own homes.[40] Although Gooday provides a rigorous discourse analysis of indoor science in the late nineteenth century, his article brushes aside the individual agency and resourcefulness of many microscopists. Ward explained that she "removed the microscope tube from the stand, and mounted it ... upon a cushion raised on a large book, so that [she] could look as through a telescope into the wine-glass."[41] Her plan "answered exceedingly well."[42] Mary Ward often ingeniously repurposed household items to make microscopic observations.[43] Gosse may have aimed at "inculcating [his readers] into a moral and orderly appreciation of 'Nature,'" as stated by Gooday, but he had little control over how his illustrations were used and adapted to someone else's observational practice.[44]

The circulation of illustrations taken from Gosse's *Evenings at the Microscope* tells another story of a microscopist attuning Gosse's observations to her own pedagogical practice. In the introduction to his *Evenings at the Microscope*, Gosse assures his readers that "the staple of the work ...

---

[38] The book was privately published in 1857 *as Sketches with the Microscope*. In 1858 a second edition appeared as *The World of Wonders as Revealed by the Microscope* and in 1864 the title was changed to *Microscope Teachings*. Overall, at least seven editions were published.

[39] An illustration included in an earlier article written by Mary Ward for the *Intellectual Observer* shows larvae inside a wineglass. Mary Ward, "A Windfall for the Microscope," *The Intellectual Observer: Review of Natural History, Microscopic Research, and Recreative Science* 5 (1864): 13–17.

[40] Graeme Gooday, "'Nature' in the Laboratory: Domestication and Discipline with the Microscope in Victorian Life Science," *The British Journal for the History of Science* 24, no. 3 (1991): 307–41.

[41] Mary Ward, *Microscope Teachings: Descriptions of Various Objects of Especial Interest and Beauty Adapted for Microscopic Observation* (London: Groombridge and Sons, 1866), 142.

[42] Ward, *Microscope Teachings*, 142.

[43] Harry writes that Ward also appears to have made microscope slides out of glass shreds. Harry, "The Hon. Mrs Ward."

[44] Gooday, "'Nature' in the Laboratory," 320.

consists of original observation" and that most illustrations were "drawn on the wood direct from the microscope."[45] By producing woodcuts himself, Gosse made sure that his book would contain largely unaltered prints of his own sketches. However, this also meant that the book's illustrations were rather coarse—at least in comparison with the illustrations produced by a trained lithographer that would later accompany his *The Rotifera*— and, perhaps more important, the illustrations were reversed left to right. This made his illustrations less useful for classificatory purposes, yet cheap and easy to reproduce. It is no surprise, then, that illustrations of rotifers included in Gosse's *Evenings at the Microscope*, aimed at adult beginners in microscopy, came to feature in an American children's book on micro- scopy, *In Brook and Bayou* (1897) by Clara Kern Bayliss (Figure 2.2).[46]

Bayliss (1848–1948) was co-editor of the progressive *The Child- Study Monthly*, which encouraged child-centered pedagogy and provided a publication for the child study movement of the 1890s.[47] Bayliss oversaw a section called the "Educational Current," in which she regularly com- mented on developments in education and reviewed books and articles. Many proponents of child studies, like G. Stanley Hall, the American leader of the child study movement, advocated for quantitative scientific methods to better understand the workings of the child's mind and improve the educational system. Bayliss's comments in the Educational Current suggest that she took a more pragmatic approach. She believed that parents needed to be educated along with their children and repeatedly argued that country clubhouses should be turned into centers for lifelong learning.[48] She wrote several children's books, which often blurred the line between fact and fiction and gave ample room to her readers' imagination. As Bayliss explained in the Educational Current of May 1899, a child's made-up stories should be seen as "fiction in its earliest and crudest form, poems and novels by untrained hands" and be encouraged, not prohibited.[49]

Philip Henry Gosse, in his *Evenings at the Microscope* had urged his readers to "[verify] ... the observations here detailed."[50] Bayliss, however, argued that the illustrations in her *In Brook and Bayou*, many of which had

---

[45] Philip Henry Gosse, *Evenings at the Microscope*, 4–6.

[46] It is not clear if Gosse's illustrations were copied by hand or if his woodblocks were purchased to be used for Bayliss's book.

[47] Emily S. Davidson and Ludy T. Benjamin, Jr., "A History of the Child Study Movement in America," in *Historical Foundations of Educational Psychology*, ed. John A. Glover and Royce R. Ronning (New York: Springer Science+Business Media, 1987), 41–60.

[48] Clara Kern Bayliss, "The Educational Current," *The Child-Study Monthly* 4 (1899): 425.

[49] Clara Kern Bayliss, "The Educational Current," *The Child-Study Monthly* 5 (1899): 49.

[50] Philip Henry Gosse, *Evenings at the Microscope*, 4.

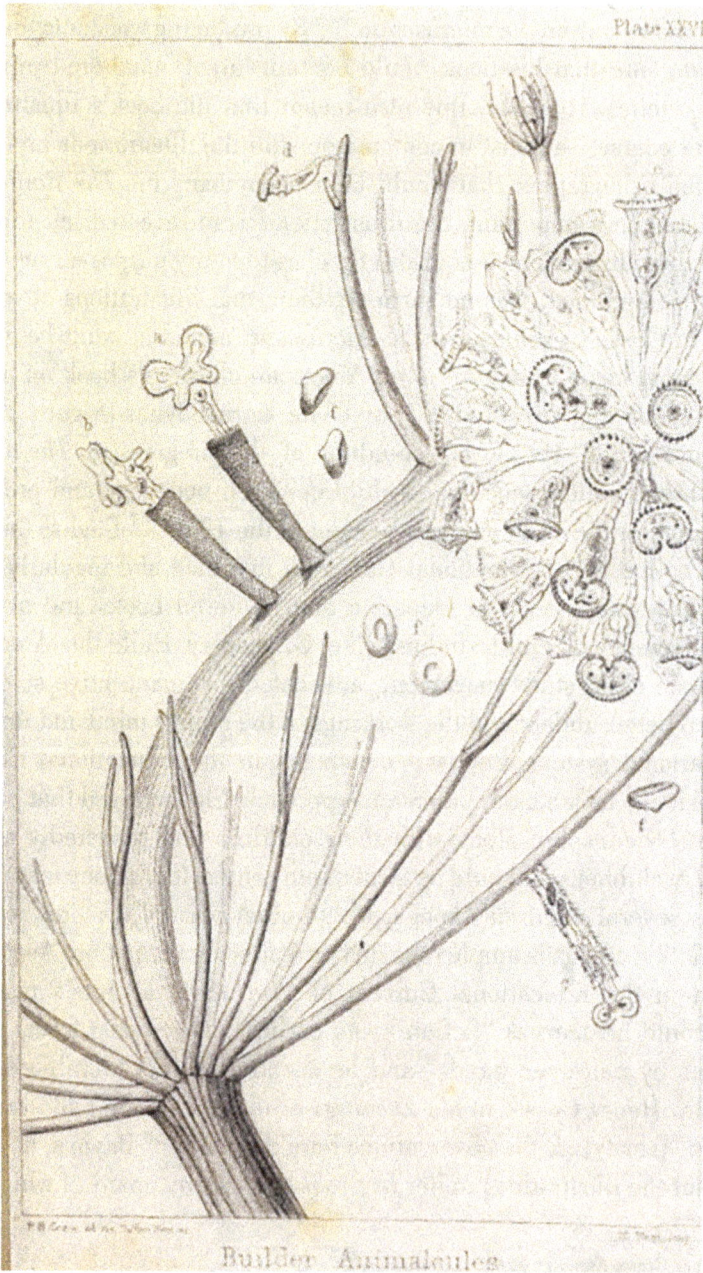

Figure 2.2 Philip Henry Gosse's illustration of the crown animalcule (a and b) and other builder animalcules in the *Popular Science Review*.

*Popular Science Review* (1862), 1, Plate XXVI. Image from the Biodiversity Heritage Library. Contributed by the Natural History Museum Library, London.

been copied from Gosse's book, "[rendered] a microscope unnecessary."[51] Instead of teaching children how and what to see through the microscope, Bayliss took her young readers on a journey into a microcosm inhabited by minute animals, mermaids, and a boy shrunk to microscopic size. Inserted into Bayliss's book, Gosse's illustrations invited the reader to explore a spectacular microscopic world, half real and half imagined. In her preface, Bayliss explained that she had written her book not only "for the purpose of enriching the child's life" but mainly "to please herself, and because she [was] fond of these microscopic creatures."[52] Evidently, *In Brook and Bayou* was not much concerned with validating observations, or introducing its readers to a community of scientific observers, and Gosse was no longer credited for his illustrations.

*In Brook and Bayou* was one of Appleton's Home Reading Books, a series meant to "extend education beyond the school."[53] However, a full-text search of the Internet Archive shows that the title of the book appears in many school library catalogues compiled around 1900. In fact, *The Elementary School Teacher* recommended use of Bayliss's book alongside magnifying glasses and pocket microscopes on school excursions.[54] Therefore, it seems probable that the illustrations in *In Brook and Bayou* were sometimes used as intended by Gosse—as a means of authenticating both the illustrator's and the reader's observations. Just like the unforeseen travels of Gosse's illustrations, the use of *In Brook and Bayou* as a school textbook and guide to excursions had not been anticipated by its author. Thus, digital full-text search can help us to shift attention away from the intended use of artifacts to their actual contexts of use.

Although Gosse's works were devoid of imaginary creatures, some of his more elaborate plates did appeal to his readers' imagination. In 1862, Gosse began to contribute a series of articles on rotifers to the *Popular Science Review*, a journal aiming to introduce its readers to a broad range of scientific fields through educational and recreational articles.[55] A footnote added to Gosse's first contribution, an article on the crown animalcule, explained that the author had "aimed so to simplify ... the matter as to convey intelligible information to such readers as have but little previous

[51] Clara Kern Bayliss, *In Brook and Bayou: Or, Life in the Still Waters* (New York: D. Appleton and Company, 1897), 12.

[52] Bayliss, *In Brook and Bayou*, 12.

[53] Bayliss, *In Brook and Bayou*, v–vi.

[54] Elsie A. Wygant, "Grade Outlines. Seventh Grade," *The Elementary School Teacher* 3 (1902–03): 456–62.

[55] For a comprehensive analysis of the contents of the *Popular Science Review*, see Ruth Barton, "Just before Nature: The Purposes of Science and the Purposes of Popularization in Some English Popular Science Journals of the 1860s," *Annals of Science* 55, no. 1 (1998): 1–33.

acquaintance with microscopical Natural History and Physiology."[56] Despite these alleged simplifications, Gosse's text catered to a specialist audience interested in rotifers. This becomes apparent when comparing Gosse's contribution with another article on "The Lowest Forms of Life" in the same volume of the *Popular Science Review*. Whereas the former provided lengthy and detailed descriptions of rotifer species, the latter, written by the editor, consisted of an entertaining dialogue between the author and a young girl looking through a microscope.[57]

However, Gosse's illustrations were more enticing than his text, offering the readers of the journal spectacular microscopic landscapes, a true "world of wonder" (see Figure 2.2).[58] Lightman has argued that such lavish illustrations were often produced with a view to instilling reverence for God's creation in their viewers.[59] Although Gosse was indeed a devout Christian, visual microscopic landscapes like his were also produced for more pragmatic reasons. As Mary Ward pointed out in her *Microscope Teachings*, microscopic panoramas served as a substitute for the actual gaze through the microscope at a time when only few could afford to buy the instrument.[60] In any case, Gosse's illustrations in the *Popular Science Review* were supposed to be marveled at, rather than to be compared with what was seen under the microscope. As I will show, Gosse's article on the crown animalcule was not the only of his publications to combine different styles of writing and illustrating, allowing its author to address different readerships at the same time. In the case of his 1862 article, Gosse's technical, written description appears to have traveled further than his illustrations.

Around 1878, the Birmingham microscopist Thomas Bolton (ca. 1830–87) established a *Microscopist's and Naturalist's Studio*, with the aim of providing microscopists with living microscopic specimens once a week. Each specimen, sent to microscopists through the post, was accompanied by a flyer consisting of a short description and illustration of the specimen. Bolton's preparations were circulated widely: They were sent to Paris, where they would then be distributed by Jules Pelletan, a French physician and editor of the *Journal de Micrographie*, and they were later advertised in the *American Monthly Microscopical Journal*.[61] The flyer issued on June

---

[56] Philip Henry Gosse, "The Crown Animalcule," *Popular Science Review* 1 (1862): 26.

[57] Lightman has argued that this "dialogue format" was characteristic of nineteenth-century popular science writing. Bernard Lightman, *Victorian Popularizers of Science: Designing Nature for New Audiences* (Chicago: University of Chicago Press, 2009), 77.

[58] Philip Henry Gosse, *Evenings at the Microscope*, 3.

[59] Lightman, "Visual Theology."

[60] According to Ward, this had changed by the time her *Microscope Teachings* (1866) appeared, which offered technical advice for using microscopes rather than microscopic panoramas.

[61] "Science-Gossip," 14–16; Jules Pelletan, "Revue," *Journal de Micrographie* 5 (1881): 351–57.

4th, 1880, was an extract from Gosse's 1862 article on the crown animalcule in the *Popular Science Review*. Gosse's illustrations, however, had been replaced by an illustration made by H. E. Forrest, another Birmingham-based microscopist, who illustrated many of Bolton's flyers (Figure 2.3).

Forrest's illustration shows a single crown animalcule instead of a whole landscape of microscopic animals. This was the regular format of Bolton's flyers: an extract from a book or journal, often summarized or adapted according to Bolton's own methods and observations, combined with an illustration of a single specimen. Bolton's lists of subscribers show that many of his specimens were sent to science educators and educational institutions—university professors, schoolteachers, museums, and colleges.[62] Bolton's flyers were mainly meant to teach aspiring zoologists and botanists how to microscopically observe and describe microorganisms. Considering that Gosse's extravagant illustrations in the *Popular Science Review* functioned as a replacement of, rather than a complement to, the gaze through the microscope, this may have been a reason to use Forrest's more descriptive illustration of a single crown animalcule instead.

Bolton's flyers were produced through autographic printing, a novel printing technique invented by the Birmingham photographer A. Pumphrey, which allowed Bolton to duplicate his flyers without the help of an engraver or printer. Instead, drawings were reproduced through a chemical process: A drawing was laid on a slab of slate coated with a kind of gelatin and a solution of bichromate of potash. The drawing ink reacted with the bichromate of potash and hardened the gelatin along the lines of the drawing, which could then be coated with ink and printed. *The Midland Naturalist* reported on a demonstration of the process at a meeting of the Birmingham Natural History and Microscopical Society and invited its readers to visit Bolton's studio and see the process firsthand.[63] Autographic printing allowed Bolton to duplicate handwritten texts and illustrations almost effort-lessly—and laterally correct. Having promised his subscribers to provide them with *living* specimens, Bolton had to describe, illustrate, and dispatch his specimens fast, which impacted his illustrations. The *Journal of Science* noticed that some of Bolton's illustrations "were roughly printed ... and dispatched hastily with some specimen unable to bear delay," but overall the journal considered the flyers "extremely well executed" and "a useful collection for reference."[64]

Some of the illustrations copied had been especially made by Forrest, but many others were copied by hand from illustrations published else-

---

[62] These lists are not exhaustive, as they only include subscribers whom Bolton deemed important to mention. However, they clearly show the educational role Bolton ascribed to his agency. Bolton, *Living Objects*.

[63] William B. Grove, "Autographic Printing," *The Midland Naturalist* 1 (1878): 132–33.

[64] "Analyses of Books," *Journal of Science* 19 (1882): 685.

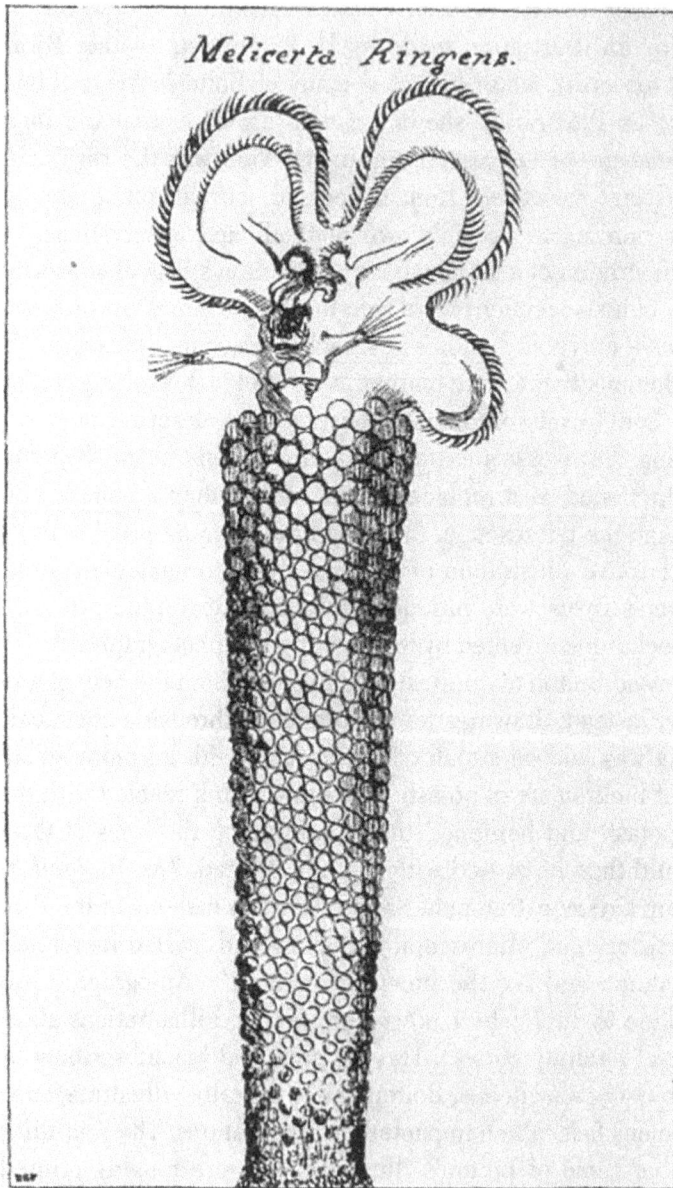

Figure 2.3 H. E. Forrest's illustration of a crown animalcule in Thomas Bolton's flyers, 1880.

Bolton, T. (1879–1882). *Hints on the Preservation of Living Objects and Their Examination Under the Microscope.* Birmingham: Herald Printing Offices. Image from the Biodiversity Heritage Library. Contributed by the Thomas Fisher Rare Book Library, University of Toronto.

where. Bolton's flyers recombined texts and illustrations taken from various journals, most acknowledging the work of other authors and illustrators by referring the reader to their publications. Bolton introduced aspiring botanists and zoologists not only to microscopical methods but also to a community of scientific authors and illustrators, occasionally distributing letters his agency had received from established microscopists. Moreover, his flyers gave the impression of being personal correspondence, as the autographic printing kept Bolton's sketchy handwriting intact (Figure 2.4).[65] Autographic printing made it possible for Bolton to increase his reach among microscopists without sacrificing the benefits of personal correspondence, like short delivery times of specimens and sharing illustrations without interference by a printer or publisher. However, the informality of Bolton's distribution network limited his authority on scientific questions. Although bound and published at intervals, not everyone considered the flyers proper scientific publications. Therefore, Bolton's discovery and description of a new rotifer species in his flyers, *Floscularia mutabilis*, was contested.[66]

A flyer issued some time in 1879 provided Bolton's subscribers with information on *Hydatina senta*, a fairly common rotifer species. The flyer combined a text previously published in Charles Dickens's *Household Words* magazine with an illustration taken from a paper presented to the Royal Microscopical Society by Charles Thomas Hudson (1828–1903), a science schoolteacher in Bristol. Hudson's illustration, unlike the decidedly low-brow text Bolton added to it, offered its viewers a close analysis of the internal organs of *Hydatina senta*.[67] Hudson had researched rotifers at least since the mid-1860s. In 1879, Edwin Ray Lankester, zoology professor at University College London, advised Hudson to join forces with Gosse.[68] The most marked result of the ensuing collaboration was the publication of *The Rotifera* in 1886, which would become a standard work on rotifers. In the years leading up to the publication of *The Rotifera*, Gosse and Hudson produced several hundred illustrations of rotifers that were to be included in their book, often reusing compositional elements of their earlier plates or whole illustrations. Notably, *The Rotifera* was not only the product of a close collaboration among microscopists, but also a visual intervention and attempted redefinition of the field of rotifer study.

---

[65] As the *Journal de Micrographie* observed, Bolton, being "l'homme d'Europe qui écrit le plus mal," decided to hire a scribe around 1881. Pelletan, "Revue." Still, his flyers continued to be handwritten.
[66] Harry K. Harring, *Synopsis of the Rotatoria*, Bulletin of the United States National Museum (Washington, DC: Government Printing Office, 1913).
[67] Bolton, *Living Objects*.
[68] Edmund Gosse, *Life of Philip Henry Gosse*, 318.

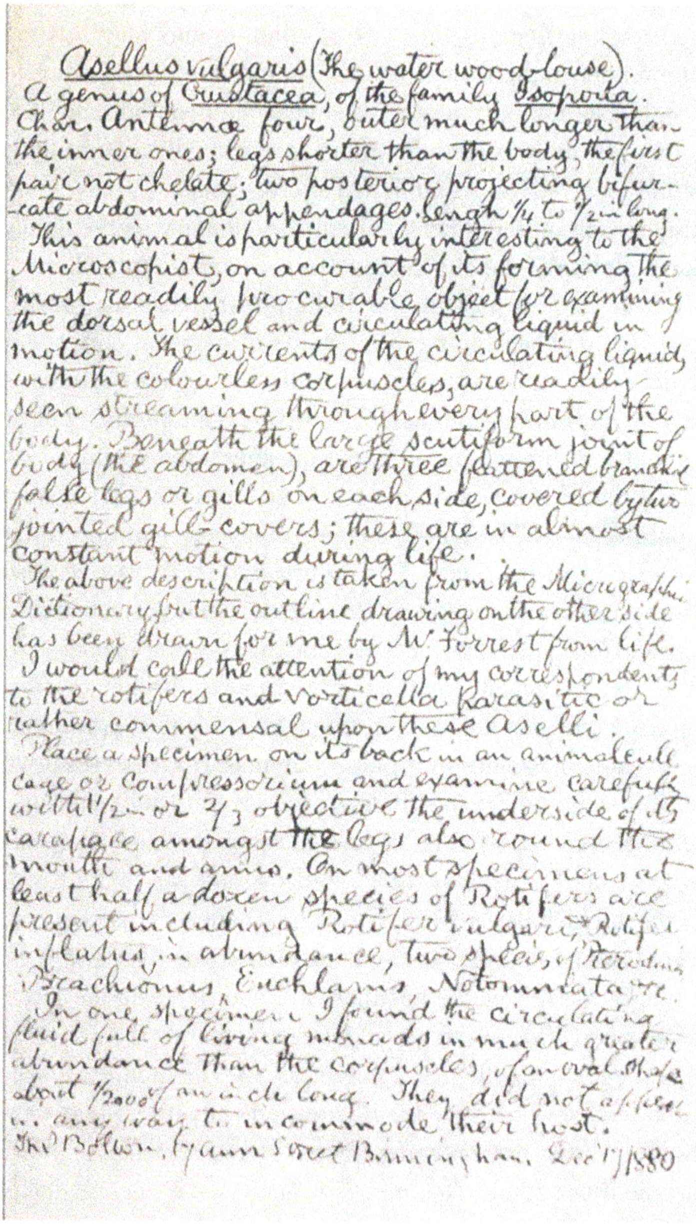

Figure 2.4 One of Thomas Bolton's handwritten descriptions, duplicated through autographic printing, 1880.

Bolton, T. (1879–82). *Hints on the Preservation of Living Objects and Their Examination Under the Microscope*. Birmingham: Herald Printing Offices. Image from the Biodiversity Heritage Library. Contributed by the Thomas Fisher Rare Book Library, University of Toronto.

In *A Naturalist's Sojourn in Jamaica* (1851), Gosse had declared natural history "a science of dead things; a *necrology*" that should be replaced with "*zoology*, i. e. the science of *living* creatures."[69] Gosse's dedication to the study of living organisms shaped his observational and drawing practices. He dated and signed the many rotifer sketches he produced for *The Rotifera* with "P.H.G. ad nat." and in the book itself, both authors asserted that all illustrations were "drawn from life." The illustrations published in *The Rotifera*, showing the developmental stages of many rotifer species, indeed testify to the large amount of time Gosse and Hudson spent observing living rotifers (Figure 2.5). Yet the illustrations were, of course, highly mediated: The execution of a drawing depended on the microscope Gosse and Hudson chose to use, their drawing instruments, and the number and quality of observations (or previous illustrations) translated into one drawing. Gosse himself occasionally included comments on the quality of his instruments, observations, and sketches in his notes.[70] And unlike Gosse's *Evenings at the Microscope* or Bolton's flyers, *The Rotifera* included colored lithographs, which required the authors to collaborate with a lithographer.[71]

Still, framing these illustrations as drawn from life and emphasizing the morphological development of rotifers served the rhetorical purpose of promoting a turn toward zoology. Their zoological turn allowed Gosse and Hudson to fashion themselves as field workers, who arguably experienced nature firsthand, and thus distinguish themselves from their metropolitan competitors who worked with collections rather than living specimens.[72] Moreover, Gosse and Hudson's rhetorical move coincided with a broader redefinition of zoology in the mid-nineteenth century. As Nyhart has shown for the German lands, zoologists, in an attempt to challenge the increasing authority of physiologists, tried to sever the ties between their emerging discipline and its predecessor, natural history.[73] At the same time, however, zoologists appropriated research objectives and practices that had been closely associated with natural history, for instance, life history studies.[74]

---

[69] Philip Henry Gosse, *A Naturalist's Sojourn in Jamaica* (London: Longman, Brown, Green, and Longmans, 1851), v–vii. See also Gooday, "'Nature' in the Laboratory," 312.

[70] Philip Henry Gosse, "135 Drawings of Rotifera by P. H. Gosse," RMS Manuscripts, Box 19, Oxford History of Science Museum.

[71] Notably, Gosse sometimes instructed the lithographer to change the size and color of his drawings. Gosse, "135 Drawings of Rotifera."

[72] Gooday, "'Nature' in the Laboratory."

[73] Lynn K. Nyhart, *Biology Takes Form: Animal Morphology and the German Universities, 1800–1900* (Chicago: University of Chicago Press, 1995).

[74] "Natural History and the 'New' Biology," in *Cultures of Natural History*, eds. Nicholas Jardine, James Secord, and Emma C. Spary (Cambridge: Cambridge University Press, 1996), 426–41.

Figure 2.5 Plate from *The Rotifera*, illustrating two genera of rotifers (*Pompholyx*, *Brachionus*) and their developmental stages.

Hudson, C. T., and P. H. Gosse. 1886/1889. *The Rotifera; or Wheel-Animalcules*. London: Longmans, Green, and Co. Plate XXVII. Image from the Biodiversity Heritage Library. Contributed by Smithsonian Libraries.

This may have been a reason why *The Rotifera*, written by a gentleman science writer and a schoolteacher from rather provincial towns in Britain, was well received overall by German zoologists.[75] Friedrich Blochmann, professor of zoology at Heidelberg University, anticipated in 1886 that the many illustrations in *The Rotifera* would be particularly useful for closing gaps in the research on rotifers.[76] After its publication, *The Rotifera* was extensively cited in the *Zeitschrift für wissenschaftliche Zoologie*, the mouthpiece of zoologists in the German lands since 1848.[77]

Alongside highly technical descriptions of rotifer species, *The Rotifera* continued a style of writing usually associated with more popular works on natural history, offering its readers entertaining accounts of the authors' country rambles. In the introduction, Gosse and Hudson reminisce about a visit to an old pond near Clifton and exclaimed, "if … we could shrink into living atoms and plunge under the water, of what a world of wonders should we then form part!"[78] The familiar tone chosen by Gosse and Hudson resonated with Jules Pelletan, editor of the French *Journal de Micrographie*. In a review published in his journal, he recommended *The Rotifera* to "amateurs" or "micrographes," whose work he considered complementary to the research undertaken by "microscopistes." As Pelletan saw it, the latter group, mainly consisting of emerging bacteriologists, treated the microscope as merely a means to an end, whereas the micrographes' enthusiasm for the microscope did more to improve the instrument.[79] *The Rotifera* thus appealed to multiple readerships, ranging from German university professors to French micrographes. With freshwater biology still in its infancy and gathering vastly different practitioners, the eclectic style of *The Rotifera* seems to have been an asset.[80]

Back in Britain, Gosse and Hudson's illustrations published in *The Rotifera* continued to undergo transformations. In the mid-1880s, Hudson

[75] Desmond has argued that for the members of the British X Club, common research agendas were often more crucial in forming scientific communities than the professional status of their members. The case of *The Rotifera* (1886/1889) suggests that Desmond's argument may also hold true for international scientific exchanges. Adrian Desmond, "Redefining the X Axis: "Professionals," "Amateurs" and the Making of Mid-Victorian Biology: A Progress Report," *Journal of the History of Biology* 34, no. 1 (2001): 3–50.

[76] "[Zu] hoffen ist, dass das in Aussicht stehende umfangreiche Werk über Räderthiere von Hudson und Gosse einen grossen Theil der noch bestehenden Lücken ausfüllen und vor Allem auch die für viele Arten noch fehlenden ausreichenden Abbildungen bringen wird." Friedrich Blochmann and Oskar Kirchner, *Die Mikroskopische Pflanzen- Und Thierwelt Des Süsswassers*, vol. Theil II. Die mikroskopische Thierwelt des Süsswassers (Braunschweig: Verlag von Gebrüder Haering, 1886), iv.

[77] Nyhart, *Biology Takes Form*.

[78] Charles Thomas Hudson and Philip Henry Gosse, *The Rotifera; or Wheel-Animalcules*, 2 vols. (London: Longmans, Green, and Co., 1886/1889), 3.

[79] Jules Pelletan, "Revue," *Journal de Micrographie* 13, no. 8 (1889): 225–30; "Bibliographie," *Journal de Micrographie* 10, no. 2 (1886): 93–99.

[80] See chapter 5 in De Bont, *Stations in the Field*.

decided to turn some of his rotifer illustrations into visual teaching aids
that he could use during science lectures at public schools. As the *Journal
of the Royal Microscopical Society* recalled in Hudson's obituary, "the
outlines of the objects were indicated by means of dots and lines, cut out
of a large brown paper screen, the perforations when necessary being
covered in with colored transparencies. When illuminated from behind, a
dark-ground effect was produced, which was most effective and elegant."[81]
The effect of Hudson's transparencies shown in a dark room must have
been dramatic—and more important than their use for microscopic studies.
Anne Secord has disputed the view that eye-catching illustrations like
these, or Gosse's microscopic landscapes in the *Popular Science Review*,
were of no scientific value.[82] Rather, Hudson's transparencies should be
considered a means of introducing students to rotifer research and recruiting
future researchers. By adapting his illustrations to his teaching practices,
Hudson helped to open the microscopy community to new members.

Thomas Bolton of the Birmingham studio died in 1887. His agency,
however, soon resumed sending specimens to European microscopists and
seems to have extended its business to North American subscribers: European and American journals, like *Hardwicke's Science Gossip* and the
*American Monthly Microscopical Journal*, continued to include adverts
for Thomas Bolton's living microscopic specimens. At the request of a
correspondent, *Hardwicke's Science Gossip* confirmed in 1890 that Bolton's
sons carried on their father's business, but the journal could not ascertain
whether the flyers were still being issued.[83] A sketchbook dating to 1889,
held by the Oxford History of Science Museum and attributed to "Thomas
Bolton," contains a wealth of rotifer illustrations accurately copied from
*The Rotifera*.[84] It seems that one of Thomas Bolton's sons continued not only
his father's business but also the circulation of Gosse's rotifer illustrations.

## CONCLUSIONS

Nineteenth-century rotifer illustrations moved through publications, flyers,
and sketchbooks, and linked widely dispersed groups of microscopists. On
the one hand, the production and circulation of these illustrations depended

---

[81] Wootton remarks that Hudson's transparencies were, in fact, produced differently, with illustrations
"painted on translucent white and coloured paper, perforated in places to create highlights. Thick brown
paper is used as a mask [...]." Wootton, "Hudson Transparencies," 65.
[82] Anne Secord, "Botany on a Plate," *Isis* 93, no. 1 (2002): 28–57.
[83] "Notices to Correspondents," *Hardwicke's Science Gossip* 26 (1890): 119–20.
[84] The illustrator assigned these illustrations figure/plate numbers corresponding with those in *The Rotifera*.

on the decisions made by the illustrator, often in collaboration with the
engraver or printer. Gosse and Hudson's illustrations in *The Rotifera* were
well received by German zoologists, as the research agenda underlying the
production of these illustrations aligned with the mid-nineteenth-century
restructuring of zoology as an academic discipline in the German lands.
On the other hand, the material qualities of the illustrations themselves
and the things (or animals) involved in their production shaped their
subsequent travels: Gosse's woodcuts, which could be reproduced cheaply,
came to be included in a children's book. Likewise, the short life span of
a microorganism sent through the post required Bolton's flyers to be printed
and distributed swiftly, which impacted the quality of their illustrations.
In any case, rotifer illustrations, even after their publication in *The Rotifera*,
remained "working images." They continued to be inextricably intertwined
with the observational practices of those who (re)used them: British zoology
students used Gosse and Hudson's rotifer illustrations to learn to spot
microscopic animals in the samples sent by Thomas Bolton. In the hands
of Bolton's son, however, the illustrations themselves became objects of
observation, teaching the aspiring illustrator how to draw scientific sketches
by copying them.

Moreover, reproducing illustrations allowed scientific authors and
illustrators to appropriate them, and gradually renegotiate what should be
considered good observational practice or who could be a reliable observer.
Gosse, in his *Evenings at the Microscope*, urged beginners in microscopy
to compare their observations with his woodcuts. In Mary Ward's *Microscope
Teachings*, Gosse's rotifer illustration appeared to result from observations
made by repurposing cushions and wineglasses. Thus, with each reproduction, rotifer illustrations offered an opportunity for microscopists to
write different groups of practitioners into the microscopy community and
portray them as trustworthy scientific practitioners, often drawing on the
authority of a previous observer by reprinting their signature. However,
these attempts were not equally successful. Since Thomas Bolton lacked
proper scientific publications, his discovery of a new rotifer species was
contested. Philip Henry Gosse, too, sometimes struggled to get the recognition he felt he deserved from his scientific circles.[85]

Digital research tools made it possible to track the multiple adaptations
of Gosse and Hudson's rotifer illustrations. Of course, it must be borne in
mind that human and institutional choices concerning the design of digital

---

[85] For example, Gosse's son, Edmund, recalled a dispute between his father and the Linnaean Society,
which criticized the religious undertone of one of his papers. Edmund Gosse, *Life of Philip Henry Gosse*.

research tools and the materials archived and digitized shape the outcomes of a digital analysis of historical sources. The optical character recognition used by the Biodiversity Heritage Library and the Internet Archive is not always accurate, and it failed to find mentions of Gosse and Hudson's monograph in German publications written in Gothic letters. Still, researching historical sources digitally allows us to look at the reception and use of artifacts and to piece together biographical information about lesser known historical actors. Without full-text search, it would have been nearly impossible to reconstruct the scientific network of Thomas Bolton, a provincial microscopist with an international reputation.

# 3

# Benjamin Smith Barton's Natural History Network: Local Knowledge and Atlantic Community

## Peter C. Messer

B etween roughly 1789 when he returned from Europe, where he had
pursued a medical degree, and his death in 1815, Benjamin Smith
Barton established an extensive correspondence network. Centered
in Philadelphia it stretched east across the Atlantic to correspondents in
England, Scotland, France, Portugal, Russia, and what now are the countries
of Italy and Germany. In the United States it extended north to south from
New Hampshire, Massachusetts, New York, Virginia, South Carolina, and
Georgia, and west throughout Pennsylvania into the Ohio Territory, Tennes-
see, Kentucky, and Louisiana. Through Thomas Nuttall, who traveled on
an expedition for Barton, it extended through the Great Lakes region north
into modern Canada and then back down the Missouri and Mississippi. It
also extended to Mexico, Brazil, and to India. Participants exchanged
and debated descriptions, specimens, and theories about plants, animals,
minerals, and people, with the hope of expanding knowledge of the natural
history and, at least in the minds of its participants, improving the world.
Their efforts in this regard unfolded both through the gradual refinement
of existing information, and through more radical disruptions as the shared
interest in advancing knowledge, in this case about the unexplored and/
or undescribed areas of North America, prompted challenges to orthodoxies
and assumptions.[1] Of course, the process of building or promoting a consensus around these ideas, whether modest or radical, came with a variety of
intended and unintended consequences that extended beyond the participants' desire to further progress in "the Natural Sciences." This chapter
highlights the ways in which Barton's network contributed to the process
of disruption and consensus building and explores some of its implications
for how people understood their world and what they might do with that
knowledge.[2]

Barton's network emerged around the possibilities that he and others
believed North America offered to expand and improve knowledge about
natural history. As he explained to the Welsh zoologist Thomas Pennant,
Barton hoped to further "the prodigious progress in Natural Science in the
last twenty years" and incorporating information from the "immense portions
of the vast continent of America" that remained "to be explored."[3] This

---

[1] Caroline Winterer, *American Enlightenments: Pursuing Happiness in the Age of Reason* (New Haven, CT:
Yale University Press, 2016).

[2] In the process the text challenges the assumption of many scholars that American science was largely
static or derivative or European practices; see Pamela Regis, *Describing Early America: Bartram, Jefferson,
Crèvecour, and the Influence of Natural History* (De Kalb: Northern Illinois University Press, 1992); Kariann
Akemi Yokota, *Unbecoming British: How Revolutionary America Became a Postcolonial Nation* (New York:
Oxford University Press, 2011).

[3] Benjamin Smith Barton to Thomas Pennant, August 30, 1790; Series I. Correspondence; Violetta Delafield–
Benjamin Smith Barton Collection, American Philosophical Society; hereafter VD-BSB APS.

possibility seemed to have formed the primary attraction to Barton's correspondents both abroad and at home. One potential correspondent in Ham-
burg hoped to "have a correspondent of your talents and knowledge in
America," and sent "an exemplar of what you can expect of me in the
natural history" along with the promise that his "many friends" across
Europe "could supply what I am unable to do for you."[4] Closer to home,
Samuel Kramsh, enthused by Barton's proposed "natural history of the
United States" advised him to encourage "friends & others ... to send you
their additions, & observations" so as to create "a more complete & lasting
work."[5] One of those friends, William Peck, as Kramsh predicted, agreed
to correspond with Barton out of his "ardent wish" to promote "harmony"
among "the lovers of Science in America & particularly those who cultivate
the sciences of Nature."[6]

This desire to improve and expand knowledge about North America,
on both sides of the Atlantic, created a context in which the network
might disrupt, or at least facilitate the disruption of, an emerging scientific
consensus. Barton's interest in this possibility emerged because his network
of local correspondents complicated his effort to know, largely through
classification, North American plants. Barton's correspondents often provided him with a wealth of information, not all of it useful in a scientific
community dominated by Linnaeus's sexual system, which classified plants
according to the number and shape of the pistils and stamen in its flower.
Robert Brown, for example, sent a package of plants, including one that
locals referred to as the valuable medicinal plant *Columba root*, but apolo-
gized that because of "my entire ignorance of botany" he could provide
only a simple physical description: "The stalk grows where it bears seed
(which is in the third year) to the height of four or five feet. The root then
dies. The leaves, I am told (for I never saw them) resemble those of the
mullein and grow on the stalk opposite to each other. It has no stalk during
the first & second year."[7] Even the more formally trained correspondents
struggled with the demands of the Linnaean system. Tucker Harris had
tried to provide Barton with "some Information" concerning "the Spanish
Potato," but had "waited until the Parts of Fructification of that plant
should arrive at the most mature state, & on careful examinations at different
Periods, until Vegetation was destroyed by Frost." Unfortunately, he "never
could observe any thing more than some Small immature white seeds" that

[4] Johann Herman Ferdinand Autenrieth to Barton, October 7, 1796; VD-BSB APS.
[5] Samuel Kramsh, June 22, 1794, to Barton; VD-BSB APS.
[6] William Dandridge Peck to Barton, October 6, 1794; VD-BSB APS.
[7] Richard Brown to Barton, October 3, 1807; VD-BSB APS.

"in this Climate at least, would never have arrived at which a state of Maturity as to produce Potatoes." What he could report, however, was how "in the most striking manner that the Potatoes, (from the slips or Vines, in particular) have varied from the original Plant, by becoming White." He "cannot account in any satisfactory way for this variety of Change in the Oeconomy of this Root," noting the he was "at a loss to determine" whether it had been caused by some "particular Quality in the Soil, or from this vegetable being cultivated too long on the same Spot of Ground."[8]

Such accounts created a dilemma for Barton and encouraged him to take a somewhat disruptive view of how plants should be classified. On the one hand, the lack of specific information rendered the information largely useless when it came to identifying plants and providing knowledge about North America. On the other hand, these descriptions were the consequence of a network that could not rely on either trained botanists or on reliable access to plants at the appropriate time or point in their life cycle, making them, in all likelihood, a permanent feature of his network. As a letter from Barton to R. A. Salisbury asking whether a "Jeffersonia" plant he had sent to the botanist had "yet flowered in England" so that it might be recognized as "a new species" suggests, Barton responded to this problem by challenging the narrow scope that the Linnaean system imposed on his work. He explained to Salisbury that the "plant belongs to the order Payareracea, and must stand next to Samgriania," unfortunately, since "the number of its stamens varies" he was not convinced that his fellow botanists would accept his conclusions. Rather than accepting this probable verdict, Barton challenged its assumption, insisting that the number of stamens, an essential component of Linnaean classification, was "a matter of small importance," and proposed that the impending controversy proved "that we must look for something better than the sexual system."[9]

Barton had ideas about what might improve if not replace the Linnaean system. He laid out his views in a lecture on botany at the University of Pennsylvania. His lecture began with the critique, illustrated in Harris's letter, that according to "the present nomenclature" plants could "only be investigated at one particular Period," which meant that botanists could not "ascertain the names unless we detect them in a certain State which shall perhaps continue but for a few Days." He believed a superior system would emerge from developing "a Language by which the Habit of Plants might be accurately described," which would answer every Purpose of

---

[8] Tucker Harris, Jr., to Barton, November 25, 1801; VD-BSB APS.
[9] Barton to Richard Anthony Salisbury, November 29, 1809; VD-BSB APS.

certainty and duration & would <u>greatly contribute to Facility</u>." He therefore
urged, perhaps thinking of the information he had received from correspondents such as Brown, "that the Habit ought to be carefully studied that
we may attempt easier & more permanent Descriptions," particularly as it
related to "the Colour, Form, Stature & other Qualities" of the plant.[10]
The system that Barton envisioned, quite simply, incorporated the sort of
diffuse information about plants—color, form, and stature—that his network
could reliably provide into the process of scientific description.

Barton did not invent this alternative system of botanical classification
out of whole cloth, it reflected a growing debate among European botanists
over how to best classify plants. Barton's call for a system that incorporated
a variety of morphological characteristics embraced the ideas the "natural"
system put forward by the French botanist Antoine Laurent de Jussieu and
promoted in England by none other than his correspondent R. A. Salisbury.[11] At the head of a network exploring and collecting the vast botanical
wealth of North America, Barton could use his resources to take sides and
perhaps advance one of these systems. Thus, when he sent Jussieu the
sheets of his work on the trees and shrubs of North America, expressing
optimism that the plan of my "Specimen" meets your approbation," Barton
offered his network in support of Jussieu's scientific project. Barton closed
his letter, after asking whether Jussieu would help get "the title, and general
objects of the works announced in some of the literary journals of Paris,"
by offering to help publish "a new edition of your excellent work, the
<u>Genera Plantarum</u>" in America, and to "supply you with seeds, or dried
specimens, of many of the American Plants."[12] Barton adopted a similar
tone with Salisbury, offering to send him either seeds or living samples of
"the rare plants of the United States" so that so that he would "have an
opportunity of describing them, and having them figured in the <u>Paradisus</u>
<u>[Londinensis]</u>."[13]

Whatever language or way of knowing the network promoted, it effectively streamlined the means of knowledge transmission. In the most basic
sense, the emphasis the network placed on classification, regardless of the
system deployed, prioritized particular ways of knowing, bestowing on them
a sense of legitimacy that it implicitly denied others. In some cases, this

---

[10] "Botany," Notes—Scientific, Misc. Folder; Series II. Subject Files: Miscellaneous Notes and Manuscripts;
VD-BSB APS.

[11] Peter F. Stevens, "How to Interpret Botanical Classifications—Suggestions from History," *Bioscience* 47,
no. 4 (April 1997): 243–50; Theresa M. Kelly, *Clandestine Marriage: Botany & romantic Culture* (Baltimore:
Johns Hopkins University Press, 2012), 36–37.

[12] Barton to Antoine Laurent de Jussieu, August 5, 1809; VD-BSB APS.

[13] Barton to Salisbury, November 29, 1809; VD-BSB APS.

streamlining of knowledge transmission by privileging one way of knowing or describing a plant had obvious benefits. One correspondent, for example, proclaimed to have never heard of "the Jestis weed" about which Barton had asked. He continued, however, that the "account of the efficacy of this weed" against rattlesnake bite corresponded "exactly with the account I have received of a root growing in Georgia" that was "known among the people of the upper country of that state, by the trivial name of the rattle snakes master." He continued, indicating that the accounts of "the power of this root (prepared & administered in the same manner)" against "the poisons of the snake were too respectable to permit me to doubt the reality," suggesting it had real utility and offered to "procure & send" Barton a specimen, if he wished "to examine this root."[14] In other words, the confusing local or common names for this plant created something of an obstacle for the dissemination not only of a particular knowledge of it as a plant but knowledge of its potential benefits. To be sure, the name *rattlesnake's master* created a clear image of what the plant could accomplish, but his correspondent's inability to recognize the Jestis weed suggests that a shared name derived from Barton's formal description would remove potential confusion on that rather important point.

The effort to promote a streamlined or specialized way of knowing plants had other consequences as well. When combined with the network's emphasis on the discovery of new species, for example, it produced an odd disjuncture in which familiar plants became alien until or unless they had been properly described. Samuel Kramsh, for example, acknowledged the limits of his index, which he attributed to having "no learned friend hereabout, with whom I could communicate" so he had "sent several Collections of dried plants to Professor Shrebes at Erlangen in Germany" for his "new Edition of Linnaeus Species Plantarum." In answer, Kramsh had learned the Professor was "surprised at so many new Species still discover'd in North America."[15] The plants, quite obviously were familiar to Kramsh and the various inhabitants of the region, but yet that information, given that Kramsh sent dried leaves, carried little weight when it came to transporting knowledge across the Atlantic. The phenomenon surfaced in America as well. One of Barton's correspondents described how his "botanical expeditions" within "a semicircle 8 or 10 miles" around the not-at-all recently settled city of Savannah produced an "astonishing" quantity of both "known and new genera and species." This sentiment in part reflected

---

[14] L. Kollock to Barton, October 11, 1808; VD-BSB APS.
[15] Kramsh to Barton, June 22, 1794; VD-BSB APS.

the heightened awareness the author's botanizing produced. He often observed "some old or non-descript that I had not seen before," but it also implicitly denied legitimacy to any knowledge about those plants obtained outside scientific description.[16] This system came with obvious costs to the intellectual authority of anyone whose knowledge relied on different ways of knowing plants. As another of Barton's correspondents explained, he had "not been able to make any improvement in the little Botanical Knowledge" because he had "no treatise on Botany" without which "it is impossible to become intimate with the vegetable kingdom." Consequently, he concluded with evident frustration, all he could do was "to make inquiry of the Old Women what use they have made of our indigenous plants."[17]

The consequences of marginalizing local names had a pronounced effect on Native Americans. An essential part of the American imperial project, according to Thomas Hallock, lay in removing Indians both literally and discursively from the landscape.[18] Botany, dating back to the Colonial period, had in many ways worked against that process because of the opportunity the demands of this empirical science provided Indians to contribute their knowledge of plants.[19] That potential was not lost on the members of Barton's network. Samuel Kramsh advised Barton, particularly with regard to medicinal plants, "to get some information form the former Inhabitants of this country, I mean the Indians, as I understand they were particularly skillful in this way."[20] Barton himself saw botanical identification as a way to, after a fashion, preserve the Indian presence in the American mind. Writing to Jeremy Belknap about a new species of "Dipus," Barton remarked that if he "could discover its name in the language of the Delawares, or any other of our Indian nations, I should, probably adopt an aboriginal name for the specific description of this little animal."[21] This practice, he explained in a manuscript note to his *Flora Virginica*, "would tend to show the attention which the original inhabitants of this country have paid to the native vegetables, it would preserve, in some measure, their language, a circumstance of no small importance in the investigation of the history of these singular people."[22]

---

[16] John Brickell to Barton, August 8, 1807; VD-BSB APS.

[17] John H. Pope to Barton, September 24, 1808; VD-BSB APS.

[18] Thomas Hallock, *From the Fallen Tree: Frontier Narratives, Environmental Politics, and the Roots of a National pastoral, 1749–1826* (Chapel Hill: University of North Carolina Press, 2003), 29–55.

[19] Susan Scott Parrish, *American Curiosity: Cultures of Natural History in the Colonial British Atlantic World* (Chapel Hill: University of North Carolina Press, 2006), 215–58.

[20] Kramsh to Barton, June 22, 1794; VD-BSB APS.

[21] Barton to Jeremy Belknap, November 23, 1794; VD-BSB APS.

[22] "Notes on the *Flora Virginica*," Folder #12, Series II. Subject Files: Botanical Notes and Manuscripts; VD-BSB APS.

The demands of the network, however, ultimately undermined these lofty goals. Barton himself, he explained to Belknap, placed limits on the practice, preferring Delaware "to those of any other Indian language because the language of the Delawares, from all I can learn, has had a more extensive spread in North America than any other Indian Language."[23] A convenient and consistent scientific practice to be sure, but it nonetheless swallowed up diversity of Indian languages and cultures into a single, purported, majority language that rendered all others unnecessary at best and invisible at worst. Barton, however, found even these limited aims thwarted by the network's emphasis on both relatively rapid discovery and identification. He explained that despite his best intentions his "labours in this subject could not be extended far" because his "collection of the aboriginal names of the vegetables of the country is as yet but very slender." Consequently, what he hoped would illustrate that "history does not furnish us with an instance of a race of men in the simplest & rudest form of Society, who have made such advances in the knowledge of vegetables" became another means of discursively removing them from North America.[24]

Barton's network attempt to establish a specific way of knowing plants and animals also marginalized Africans' and African Americans' contributions to natural history. In the Colonial period, as Susan Scott Parrish has noted, African Americans had been seen, as a result of their familiarity with plants and subtropical environments, as useful collectors of botanical curiosities.[25] Barton's network accepted the idea that African Americans, particularly slaves, might serve as useful collectors, but its focus on using specific classificatory systems as a measure of knowledge minimized their role in its production. Barton noted, for example, that "William Bartram informs me that 'Good Night to you' is the name of a species of Cassia, both in N. & S. Carolina, among the negroes, the country people, & others." The name, he continued, came from the observation that as the light waned in the afternoon "the foliole" of the plant closed "upon each other, & hang down, during the night." However, even as the description acknowledged the contributions of the "negroes" and country folk in calling Bartram's, and his, attention to a plant, it also marginalized their way of knowing it. The plant's curious properties, information that tied the production of knowledge to those who could see the living plant and understood it by its human affinities, served only as a temporary marker in the quest to further differentiate among species of Cassia. In the process, the network

---

[23] Barton to Belknap, November 23, 1794; VD-BSB APS.

[24] "Notes on the *Flora Virginica*," VD-BSB APS; see Schiebinger, *Plants and Empire*, 194–225.

[25] Parrish, *American Curiosity*, 271–74.

shifted the production of knowledge away from human interactions with
the world around them, a system that privileged those, such as the "negroes"
and country folk, who lived among the plants and animals they sought to
know, to the ability to place specific parts of a plant within a specialized
classificatory system.[26] Barton's narrow description of the plant's curious
characteristics illustrates this shift. The name *Good Night to You* equated
the ability to identify the plant with the ability to recognize, or imagine,
in the plant a connection to the observer founded on a broadly shared
human sentiment that the plant's behavior evoked. Barton, however, by
explaining the name as the product of the way a very particular part of
the plant, identified by its name in the specialized language of botanical
Latin, the foliole, reacted to the waning light—made knowledge of the plant
contingent on being aware of plant morphology and not human experience
or sentiment. He completed that association by observing that Bartram
believed the plant was the "*stearis bipured* Dillenium (whose figure I showed
him) vol 1 1792" or, the "Cassia obtusisabia of Linne," a figure that almost
certainly consisted of nothing more than a drawing of the leaves and
flowers of the plant.[27] In other words, the "negroes'" contribution consisted
primarily of describing a curious quality of the plant that invited further
investigation, but that was not useful, even in Jussieu's system, unless it
could be placed in a classificatory system that they neither used nor, at
least in this description, understood.

A more direct dismissal of African American ways of knowing appeared in Barton's attempt to identify a new species of salamander, "known
in America as the Alligator or Hell-Bender." In this publication, intended
for his "friends," but "especially" his "foreign correspondents," Barton
simultaneously acknowledged and dismissed potential African American
contributions to his scientific endeavors. He explained that the animal's
colloquial name "Hell-bender" came from "the negroes in the western
parts of Virginia," effectively identifying them as a likely source of some
of the "interesting circumstances in its history" included in the memoir.
With the origin of the name established, however, he immediately offered

---

[26] On the clash of ways of knowing inherent in specialized description, see Brian W. Ogilive, *The Science
of Describing: Natural History in Renaissance Europe* (Chicago, 2006); Michel Foucault, *The Order of
Things: An Archeology of the Human Sciences* (New York: Random House, 1994, 1970).

[27] The reference to *Dillenium* is to the herbarium of Johann Jacob Dillenius (1684–1747) Sherardian
professor of botany at Oxford; see *The Literary Gazette and Journal of the Belles Lettres, Science, and Art,*
1851 (July 10, 1852), 547; "The Biographical Memoir of John James Dillenius, M.D. Sherardian Professor
of Botany at Oxford," in *A Selection of the Correspondence of Linnaeus, and Other Naturalists,* vol. 2, ed.
James Edward Smith (Cambridge: Cambridge University Press, 2014), 82–84. For an example of Dillenius's
botanical drawings see, Johanne Jacabo Dillenio, *Plantarum Rariorum Quas In Horto Suo Elthanmi in
Cantio* (London, 1732), plate 148.

an epistemological dismissal of the scientific utility of any other information coming from this source. The name came from the animal's "slow twisted motions, when moving in the waters, which the slaves compare to the tortous pangs of the damned in hell," prompting Barton to ask whether it was "beneath the dignity of natural history to notice such vulgar names, when they serve to throw any light upon the habits or economy of the animal?" In short, natural history as a science was not compatible with the way slaves understood the world, though that understanding might occasionally, and accidentally, provide some useful information, in this case an accurate, if overly colorful, description of the salamander's movements. Barton reinforced that sentiment by suggesting that the name should lead "the moralist" to ask whether there was "something melancholy and distressing in the condition and reflections of those who impose such names?"[28] On the bright side, Barton was deploying his network to encourage readers to ponder the moral evils of slavery, though in doing so he was also encouraging them to view African Americans, or slaves, as particularly ill suited to scientific inquiry.

Barton, of course, had incorporated the name into the title of the work, but even that should be seen in a mixed light. He implicitly acknowledged the name's utility in his description of the animal's "slow, tortuous movements," but he noted these were "far from being graceful, as are those of many other *reptilia*" and tended to "increase our prejudice against the animal." Perhaps eager to distance the animal from the unpleasant connotations, and, implicitly, from those who provided the name that evoked them, Barton, in the text of the publication, referred to the animal by its Delaware name, *Tweeg*, however accurate or useful Hell-bender was for conveying a sense of how the animal moved. In other words, Barton sought to distance the creature from the rigorous analysis of the form and history of the salamander, which constituted the principal subject of the memoir. Unfortunately, Barton's use of the Delaware name for the animal did not encourage his network to take their ability to contribute to natural history particularly seriously either. He admitted he preferred this name despite the fact its "precise import" he did "not know," removing any implication that the Delaware way of knowing the animal held any particular importance. A sentiment he reinforced by justifying his decision to use *Tweeg* because it had also "been adopted by the white inhabitants in certain parts of North-America."[29]

---

[28] Benjamin Smith Barton, *A Memoir Concerning an Animal of the Class of Reptilia, or Amphibia, which is Known, in the United States, by the names of Alligator and Hell-Bender* (Philadelphia: Griggs & Dickinson, 1812), iii, 5, 6.
[29] Barton, *A Memoir Concerning an Animal 6, 18.*

The manner in which Barton's network defined and produced knowledge also contributed to the marginalization of African and African American medical knowledge. The subject was of no small importance to both
planters and slaves, as the former exerted their authority over slave bodies
by subjecting them to various treatments for illnesses, while slaves resisted
by cultivating alternative medical practices intended to heal the body and
soul.[30] Barton's correspondents enlisted his network in this struggle to
present slaves as bodies that white planters and doctors could use to
produce knowledge. Charles Giguilliary, for example, hoped to use Barton's
patronage to secure an appointment as an army surgeon, which would
enable him to "experiment" with a plant, common in the Georgia "low
Country called the wild Hippo ... which the old planters tell me, is the
best thing they have tried among their negroes for the secondary symptoms of
syphillis and in Gonnorehea virulenta."[31] Similarly, John Brickell described
that although many "Women far gone in pregnancy" had "miscarried" his
"negro women have escaped abortion, by putting flannel next to the skin"
and by "keeping their bowels moderately open by castor oil, or such gentle
things: and suppressing the fury of the cough by dint of low diet and
laudanum."[32] John Casey described how the "medical virtues" of "Gossyp-
ium" as a means of treating burns had come to light by "accident." It
seems that a "negro child playing near a fire" had received serious burns,
and "his mother picking him up & finding thus cruelly burnet threw him
into a pile of callow while she ran to notify her mistress; on their return
in half an hour the child was asleep."[33] It is worth noting that in none of
these cases is the origin of the treatment entirely clear. Given the prevalence
of slave medical practitioners and their use of herbs, there is no reason to
believe that slaves themselves might not have known or discovered Hippo's
medicinal uses, or the beneficial effects of laxatives and narcotics, or even
the sedating quality of callow.[34] Framed as successful medical experiments
undertaken by White doctors, however, these letters and the network that
distributed them eliminated that possibility, while claiming for their authors
the absolute right to treat and learn from slave bodies.

[30] Rhys Isaac, *Landon Carter's Uneasy Kingdom: Revolution and Rebellion on a Virginia Plantation* New York: Oxford University Press, 2004), 116–88; more generally, see Pablo F. Gòmez, *The Experimental Caribbean: Creating Knowledge and Healing in the Early Modern Atlantic* (Chapel Hill: University of North Carolina Press, 2017); James H. Sweet, *Domingos Álvares, African Healing, and the Intellectual History of the Atlantic World* (Chapel Hill: University of North Carolina Press, 2011).
[31] Charles Giguilliary to Barton, Savannah, November 23, 1813; VD-BSB APS.
[32] John Brickell to Barton, November 25, 1787; VD-BSB APS.
[33] John M Casey to Barton, Augusta, GA, September 1, 1813; VD-BSB APS.
[34] Philip D. Morgan, *Slave Counterpoint: Black Culture in the Eighteenth-Century Chesapeake and Lowcountry* (Chapel Hill, University of North Carolina Press, 1998), 626–29.

The network's tendency to silence African and African American medical knowledge was also evident in the way members presented their treatments as effective. As Pablo Gòmez has argued, African medical knowledge was not founded on what plants or remedies treated what illnesses, but on the use of these remedies in specific ritual surrounding their application and the sense experiences of the illness, the medicine, and the recovery.[35] Moreover, James Sweet has stressed that the rituals and sensations surrounding these treatments often unfolded in a communal context; the healer applied the medicines but the movements and sounds that ultimately made them efficacious came from an assembly of friends and neighbors that constituted an essential part of the ritual.[36] These letters contained no such information in the results they communicated back to Barton. Giguilliary identified the plants' "very active properties" as the source of their effectiveness, which he believed Barton could use to good effect if provided with "the results" of Giguilliary's experiments along with "with some of the seed, and dried roots, and ... a drawing of the plant in full bloom."[37] Casey's insistence that the initial experiment was one of the "many cases of its immediate relief and successful application" that had "come within my own observation," suggests a broadly similar understanding of the process of treating illnesses and complaints.[38] Brickell provided a larger context for the problem his patients faced, if not the reasons the treatment worked, but found it in a passing "comet, which by its attraction or repulsion upon the light, heat, electric, and magnetic fluids, in the interstellar regions" might have formed "a new compound—and a new stimulus—or aid in increasing the asperity of acting or enciting causes."[39] These letters made no mention of any ritual surrounding the slaves' medical practices or experiences, and instead enlisted the network as a means of communicating straightforward empirical data that deprived any validity to the way both practitioners and patients understood medicine and treatment.[40]

The network contributed to the suppression, or obscuration, of other forms of knowledge as well. As Londa Schiebinger has argued, the late eighteenth century was a critical moment in the evolution of the medical profession's attitudes toward abortion, notable in the disappearance of any reference to drugs as potential abortifacients in the scientific medical

---

[35] Gòmez, *The Experiential Caribbean*, 95–117.

[36] Sweet, *Domingos Álvares*, 123–45; Gòmez, *The Experiential Caribbean*, 118–44.

[37] Charles Giguilliary to Barton, Savannah, November 23, 1813; VD-BSB APS.

[38] John M. Casey to Barton, Augusta, GA, September 1, 1813; VD-BSB APS.

[39] Brickell to Barton Nov. 25, 1787; VD-BSB APS.

[40] Gòmez, *The Experiential Caribbean*, 116–17.

literature.[41] Barton's network appears to have done its part to facilitate that process. James Anderson, for example, wrote to Barton with information he had collected from "a Learned Clergyman" about the "excellency of Stramonium," which offered a "Remedy for those females whose courses (or Monthly flowing) are stopped." The clergyman agreed to provide Anderson with the "the Recipe," but only after his "promising him to be communicating it to none but a botanist." The caution evidently arose from concerns that the same decoction that cured suppressed menstruation was "the most dangerous potion that can be given to pregnant women and whoever of the females use it must use it with the utmost caution." The plant is toxic, thus Anderson's warning may have been merely precautionary, but the emphasis on the danger to pregnant women suggests that it may also have been an abortifacient. A supposition made more plausible because Anderson also reported that a man used it as a "a cathartic" with no ill effect. On the contrary, according to Anderson's source, the "Gentleman" had used it to effect a "surprising cure" on, given Anderson described it as "I shall not mention," a venereal complaint.[42]

Barton would probably have welcomed Anderson's information. The 1796 Philadelphia edition of William Lewis's *The Edinburgh New Dispensatory*, one of the most influential and important pharmacopeias published in both Britain and America, called particular attention to stramonium.[43] Lewis had acknowledged the plant's deserved reputation as "a strong narcotic poison," but nonetheless contended that it "deserves the attention of practitioners, and well merits a trial, in affections often incurable by other means."[44] Given that Barton saw discovering and cataloging new medicinal uses for plants as a way to assert American equality in the sciences and enhance his own reputation, he almost certainly would have seen Anderson's information as an important opportunity to pursue.[45]

Barton appears to have embraced the opportunity that Anderson's letter offered. When he published his own version of William Cullen's *The Materia Medica* in 1811, he included among the "many additions" that he made to the text to make it "more useful to the student of medicine" the results of American experiments with stramonium.[46] The text reprinted

---

[41] Londa Schiebinger, *Plants and Empire: Colonial Bioprospecting in the Atlantic World* (Cambridge, MA: Harvard University Press, 2004), 150–93.

[42] James Anderson to Barton, August 4, 1804; VD-BSB APS.

[43] On the influence of the *Edinburgh New Dispensatory* and its introduction of stramonium, see David L. Cowen, *Pharmacopoeias and Related Literature in Britain and America, 1618–1847* (Burlington, VT: Ashgate Variorum, 2001), 70–71, 68–69.

[44] William Lewis, *The Edinburgh New Dispensatory* (Philadelphia: Dobson, 1796), 249.

[45] Cowen, *Pharmacopoeias and Related Literature in Britain and America*, 201.

[46] Benjamin Smith Barton, Editor's Preface, in William Cullen, *Professor Cullen's Treatise of The Materia Medica with Large Additions, including Many New Articles wholly omitted in the Original Work*, 2 vols. (Philadelphia: Edward Parker, 1812) I:xii, xvi.

Cullen's original observation that the plant was "a powerful narcotic substance," with the "seeds" having proved to "have been especially remarkable in this way," and that despite various trials he had "not" found any "proof of any peculiar power in the stramonium," in curing mania or epilepsy. Barton, however, added that the plant was "an article of great powers and value" as used in "the United States," noting his own experiences with the drug in treating epilepsy, mania, and "melancholia." He also added that an extract derived from the plant and its seed had "afforded essential relief" to cancer patients and, notable in light of Anderson's reference to the ailment he would not name, that it had been useful in treating "cases of rheumatism, especially syphilitic rheumatism" in cases of "ulcerous affections attended with high irritation."[47] Although stramonium's utility in treating symptoms of syphilis had made it onto Barton's list of the plant's medicinal uses, he made no mention of its potential uses in treating suppressed menstruation nor warned about the dangers it posed to pregnant women.

Of course, Barton's decision to limit the discussion of stramonium as a treatment for epilepsy, mania, melancholia, cancer, rheumatism and syphilis may reflect the genuine state of knowledge about the plant. A modern study of Ethiopian folk medicine that identifies stramonium seeds as an abortifacient, however, should caution us from assuming Barton simply omitted Anderson's warning because it was baseless.[48] A decision to either ignore or suppress that information, it should be noted, would conform to what Schiebinger and other scholars have identified as a contemporary trend among doctors and pharmacists to avoid experimenting with or identifying potential abortifacients.[49]

Viewed in the context of Barton's treatment of potential abortifacients in his version of William Cullen's *Materia Medica,* this suspicion seems even more warranted. Despite popular ambivalence about abortion, information about potential abortifacients was relatively widely available in works such as Nicholas Culpepper's *Complete Herbal.*[50] Thomas Short's *Medicina*

---

[47] Barton, ed., *Professor Cullen's Treatise of The Materia Medica,* II:200, 201.

[48] Soloman Araya, Balcha Abera, and Mirutse Giday, "Study of Plants Traditionally used in Public and Animal Health Management in Seharti Smere District, Southern Tigray, Ethiopia," *Journal of Ethnobiology and Ethnomedicine* 11, no. 22 (2015): 8. Another modern study cites, without specific reference to consequences, the dangers that stramonium poses to pregnant women: Priyanka Soni, Anees Ahmad Siddiqui, Jaya Dwivedi, and Vishal Soni, "Pharmacological Properties of *Datura stramonium L.* as a Potential Medicinal Tree: An Overview," *Asian Pacific Journal of Tropical Biomedicine* 2, no. 12 (2012): 1005.

[49] Schiebinger, *Plants and Empire,* 190–93; Edward Shorter, *History of Women's Bodies* (New York: Basic Books Inc, 1982), 181–82; Angus McLaren, *Reproductive Rituals: The Perception of Fertility in England from the Sixteenth Century to the Nineteenth Century* (London: Methuen, 1984), 110–11.

[50] Janet Farrell Brodie, *Contraception and Abortion in 19th Century America* (Ithaca, NY: Cornell University Press, 1994), 42.

*Britannica*, published in Philadelphia in 1751, listed, despite his assertions to the contrary, six plants likely to cause an abortion or kill the fetus in the womb—Gladwin Iris (*Spatula fœtida*), Ground Pine (*Chamæpitys*), White Hellebore (*Hellaborus albus*), Hyssop (*Hyssopus*), Rue (*Ruta*), and Savin (*Sabina*). Given Short's decision to cross-reference "Abortion to cause" with "Birth and After-Birth to expel" it seems reasonable to also consider another thirty-two plants listed as likely to induce uterine contractions necessary to expel a dead fetus/child, birth and afterbirth, or stimulate menstruation as abortifacients as well.[51] Barton, in keeping with Schiebinger's argument, departed from this earlier tradition and made no mention of any plants as useful in promoting birth or expelling afterbirth or a dead child, though he did list ten emmenagogues, or plants capable of stimulating menstruation. Of these, three were found on Short's list of clearly identified abortifacients, though none were acknowledged to have that power in Barton's edition of Cullen. Hyssop was useful against coughs, rue had "some emmenagogue virtues," but, added Cullen, had "not been so successful in employing them."[52] Barton, quoting Cullen, claimed that savin, perhaps the most common eighteenth-century herbal abortifacient, because of its "acrid and heating" qualities, physicians were "prevented from employing" it as an "emmenagogue"; he admitted that the plant showed a "powerful determination to the uterus," but immediately reminded readers that despite that property Cullen had "been frequently disappointed" in his efforts to use it as an emmenagogue.[53] Barton did list several other noted herbal abortifacients but minimized their effectiveness as emmenagogues. Pennyroyal (*Pulegium*) did "not assist in menstrual evacuations," and suggestions to the contrary were made without "any discernment"; sage was merely an "aromatic"; Cullen had been "disappointed" in saffron's effectiveness as an emmenagogue; tansy (*Tanacetum*) was useful for the treatment of gout; and jalap (*Podophyllum*) was a purgative.[54] Barton did list several drugs as effective emmenagogues that were also known abortifacients, but nothing in his description suggested their latter properties. Rattlesnake root (*Polygala senega*), "recommended as a

---

[51] Thomas Short, *Medicina Britannica* (Philadelphia: Franklin & Hall, 1751), on his decision not to describe plants likely to cause an abortion, viii; the abortifacient qualities of the six plants, 119, 128, 134, 152, 247, 255; his cross-referencing of abortion and birth and afterbirth, 314.

[52] Barton, ed., *Professor Cullen's Treatise of The Materia Medica*, 108, 258.

[53] On savin's notoriety as an abortifacient in the eighteenth century, see, Shorter, *History of Women's Bodies*, 186–87; McLaren, *Reproductive Rituals*, 104–05. Barton, ed., *Professor Cullen's Treatise of The Materia Medica*.

[54] On the notoriety of pennyroyal, sage, saffron, tansy, and jalap, see Shorter, *History of Women's Bodies*, 180, 186–88; McLaren, *Reproductive Rituals*, 104–05; Brodie, *Contraception and Abortion in 19th Century America*, 42–43. Barton, ed., *Professor Cullen's Treatise of the Materia Medica*, 107–09, 222, 57, 376.

valuable menagogue"; common madder (*Rubia tinctorum*), showed "considerable effects on the uterus"; whereas the "emmenagogue quality" of rosemary (*rosmarinus*) had "been observed by some of the physicians of Philadelphia."[55] Barton, in other words, appears, at the very least, to have avoided using his network to spread knowledge about potential abortifacients, and perhaps, in light of the omissions regarding stramonium and other herbs, used it to suppress, at least among scientifically inclined physicians, that knowledge.

The knowledge Barton's network produced had political as well as scientific and social implications. The exchange of scientific information, for example, could help establish or normalize political boundaries and assert national sovereignty both abroad and at home. In some ways the network, with its emphasis on promoting knowledge, worked against establishing strong national identities. As Barton explained to Pennant, while he was "very partial" to his own country, he declared he was "not, however, so partial, that I cannot love other countries and love or venerate their good and great men." Yet these assertions of comity and exchange unfolded against a rather fraught political backdrop. Barton indicated he was willing to maintain his correspondence with Pennant despite Britain's attempts "to deprive America of the greatest gift heaven has bestowed on man," explaining "I love not Politics. My happiness is wrapt up in other things. The volume of nature is my object."[56] These sentiments present Barton as a member of the enlightened generation of transatlantic scientists, but it also represents a subtle reworking of the relationship between collectors and observers in America and those in Europe. During the Colonial period, the absence of a political boundary between the Americas and Britain and the transfer of scientific information from one to another represented an affirmation of an imperial relationship in which colonists worked to promote imperial ends. American independence, however, disrupted that relationship by raising the possibility that political concerns might justify or require a suspension of intellectual exchange. Barton's insistence that he was above such things, therefore, was a notable gesture in support of free scientific inquiry, but also an assertion of national sovereignty as he committed himself to an intellectual exchange that, unlike in the Colonial relationship, his correspondents could not simply assume he would make.

Of course, such assertions would not necessarily change the reality of a relationship in which American collectors provided specimens and

---

[55] On the less common use of these herbs as abortifacients, see Brodie, *Contraception and Abortion in 19th Century America*, 43–44. Barton, ed., *Professor Cullen's Treatise of the Materia Medica*, 414.
[56] Barton to Pennant, April 7, 1793; VD-BSB APS.

drawings to Europeans who did the real scientific work. Barton's acknowledgment that his fellow citizens were "as yet, infants in natural History"
seemed to point in that direction, but he attempted to frame the knowledge
that circulated through his network in a very different way. He, for example,
argued that recognizing Europe as the publishing and distribution center
for information on natural history was merely a first step in encouraging
American intellectual independence. He was willing to contribute to Pennant's "important work," his *Artic Zoology*, because it would "more especially" serve "the interests the lovers of natural science in the New World,"
as "the blessings of peace and harmony" had "introduced a taste for this,
as well as for other sciences."[57] He also used his network to highlight his
own original contributions to natural history. In his letter to Salisbury
promising information about American plants, Barton recommended his
proposed *Flora* as it would "contain many new plants and what I am sure
will please you still more, an extension of our knowledge of species already
named." Information that would help Salisbury with his efforts to establish
and promote a natural system of plant classification. Barton also highlighted
his potential contributions to the emerging field of plant geography.[58]
He explained that his *Flora* would reveal that American "mountains and
especially the northern part of the United States" were "rich in the plants
of Asia and Europe," particularly "the northwest" in which his informants
had described "many of the Plants of Eastern Asia. No botanist more than
yourself knows how to value such facts as these."[59]

For the most part, members of Barton's network seemed willing to
accept his various conflations of national sovereignty and scientific exchange. Some embraced the visions of sovereignty implied in Barton's
exchange out of genuine sympathy for the American political experiment.
Writing from Scotland, Samuel Paterson explained his interest in corresponding with Barton because "Freedom or Liberty is such a rare thing
Among the Nations of the World that every lover of Mankind is bound to
Encourage a Nation where it is."[60] A similar sentiment came of John
Lettsom, who claimed that at a time when "many men of Science are
discouraged from the chaos around them the cool reception" toward their
ideas, there appeared "so much good sense in America, and such powerful
means in its inhabitants" that "any useful plan" to promote scientific study

[57] BSB to Thomas Pennant, in London, August 30, 1790; VD-BSB APS.
[58] A. G. Morton, *The History of Botanical Science: An Account of the Development from Ancient Times to
the Present Day* (New York: Academic Press, 1981), 313–14.
[59] Barton to Salisbury, November 29, 1809; VD-BSB APS.
[60] Samuel Patterson to Barton, Edinburgh, June 11 1808; VD-BSB APS.

"will ultimately succeed."[61] Pennant framed his interest in correspondence in more pragmatic terms. He initially accepted Barton's offer of correspondence: because he wished "to usher a rising genius of the New World to the literati of the old," implicitly recognizing Barton's current and potential contributions to a broader scientific project in exchange for a continued supply of specimens.[62] Subsequent correspondence added a further layer of pragmatism more tied to political realities. The Welshman informed Barton that he would "be at all times happy to hear of your welfare & that of your country for one reason in particular that the prosperity of one country depends much on that of another"; a sentiment that reflected in no small part his subsequent admission that the ongoing conflict "with France has most sensibly injured us."[63]

Barton's network not only helped secure recognition of American sovereignty abroad, it also helped secure sovereign authority at home. Thomas Jefferson, for example, enlisted Barton's help in preparing Merriweather Lewis for the expedition to explore the Missouri and "whatever river, heading with that, runs into the Western ocean." Jefferson appealed to their shared "wish to promote science" to persuade Barton to instruct Lewis in the subjects "in the lines of botany, zoology, or Indian history, which you think most worthy of enquiry & observation."[64] A few years later Jefferson made a related request, sending for Barton's "investigation" several "dried specimens of plants" that "Mr. Dunbar" had collected during his excursion up the Washita"; Jefferson added that Dunbar's "journal & Dr. Hunter's, furnish us with the geography of the river, accurately taken and with a good deal of matter relative to its natural history."[65] Barton himself embraced a variety of opportunities to integrate scientific inquiry into both veiled and explicit projects of continental expansion. He attempted to persuade Secretary of the Treasury Albert Gallatin to provide Thomas Nuttall with a passport for his expedition through the northwest and Canada by noting that he would be collecting information on "plants and other objects of natural history," as well "information relative to the manners, customs, &c of the Indians," that Barton required before publishing "some extensive works relative to our country."[66] Barton also took advantage of the patronage relationships that he established with young men whom he

[61] John Coakley Lettsom, no date or place; VD-BSB APS.

[62] Pennant to Barton, October 17, 1790; Box 1: Correspondence; Benjamin Smith Barton Papers, Historical Society of Pennsylvania; hereafter BSBP HSP.

[63] Pennant to Barton, August 1793, BSBP HSP.

[64] Thomas Jefferson to Barton, February 27, 1803; BSBP HSP.

[65] Jefferson to Barton, May 2, 1805; BSBP HSP.

[66] Barton to Albert Gallatin, March 14, 1810; VD-BSB APS.

recommended for jobs as surgeons in the army. One correspondent writing for Barton's recommendation for a position in the Army asked for information so that he might use the opportunity presented to "avail" himself "of every opportunity of contributing my mite to the advancement of science"; sentiments echoed by another correspondent whose successful appointment as an army surgeon led him to promise "to collect for you specimens of any animals or plant high I should think curious or useful."[67]

The implications of these expeditions were both obvious and subtle. The routes the men traveled, their observations about the native peoples, their experiences with the landscape, and profitable discoveries all provided potentially valuable information to settlers or armies moving west. Thomas Barton, for example, while serving in Alabama reported back that he had found "as few minerals as in any ... district or country in the United States," but also noted the "salt spring in the Tombigbee" was "the only quarter" from which the mineral was found in the country. He also noted that the ridge of limestone might provide a "substitute "for the French bass stone." He also reported a root that produced "a beautiful red" red dye, "not inferior to the Turkish red" for which the "Indian name is Push-shi-yea," and when combined with the yellow dye from the laurel "cannot be washed out."[68] Even seemingly smaller and more explicit scientific contributions could play a similarly expansionist role. Jefferson, for example, sent Barton "drawings & specimens of the seed, cotton, & leaf of the cotton tree of the Western country," prompting Barton to respond that the tree was "not known to the generality of botanists," having only been "briefly described by Marshall," but noting that it was "native of many parts of the country that is watered by the Ohio, the Missouri, the Mississippi, and other great rivers."[69] On its own, the report is a relatively innocuous description of an interesting tree. In the context of Jefferson's observation that the tree's boughs, according to "the journals of Lewis & Clarke," were "the sole food of the horses up the Missouri during winter," however, it also becomes an almost proprietary piece of information necessary to move across the continent.[70]

A slightly more abstract phenomenon appears in the way the network's scientific description normalized and naturalized the incorporation of Indian lands into the United States. As Catherine Tatiana Dunlop has argued,

---

[67] J. Rice to Barton, July 11, 1812; Charles Giguilliary to Barton, November 23, 1813; VD-BSB APS.

[68] Thomas W. Barton to Barton, April 2, 1814; VD-BSB APS.

[69] Jefferson to Barton, December 22, 1805; BSBP HSP; BSB: Notes—Scientific, Folder 1; Series II: Subject Files; Miscellaneous Notes and Manuscripts; VD-BSB APS.

[70] Jefferson to Barton, December 22, 1805; BSBP HSP.

identifying shared species of plants, along with soil types and topography, provided a natural and scientific basis for drawing geopolitical lines to bring once separate and sovereign jurisdictions together into one geopolitical entity.[71] Viewed in this light, Milton Anthony's letter to Barton informing him that he had found "Polygala Lemegox D. Cox inforsius," a useful medicinal plant, in northwest Georgia would have done a great deal to encourage Barton and others to see the lands "lately obtained from the Cherokee Indians & added to this State" as a logical extension of their nation's boundaries. When he added that the plant grew there in "greater quantities ... probably than any other part of the United States," he would have added a degree of necessity or utility to the logic of adding this new territory. Finally, his report that the area also contained a great variety of "native plants which would afford you great pleasure if convenient for you to make an excursion through that country" provided a new set of biological affinities, rooted in their connection to the "Polygala Lemegox D. Cox inforsius," that might justify further expansion.[72]

In the broader context of scientific exploration, the network applied the same implied logic to scientific questions as well as known scientific facts. Thomas Barton, Benjamin's nephew, while serving in Alabama, reported back that he found "Petricfications" of "sea shells" on "a limestone bluff which runs through this country in a direction with the Allegheny." He also informed his uncle that he had "collected a fine seed for you some of which may be new, and that the "Herbaceous plants" were "in full vegetation, and had I means of collecting and conveying to you; or did I know the name so as to describe them, then I could I have no doubt give you much pleasing information on them." He also reported that the interpreters in camp from the "Chackaws, Chickasaw's, and Creeks" knew "nothing of a bird of the name you have given wither in English or Indian," but from Barton's description they reported, "that in the neighborhood of Pensacola there are many of them." He also noted that "Swallows and Martins appear there about the month of February."[73] Prompted by Barton's questions, Thomas had transformed Alabama into something of laboratory from which answers to larger scientific questions might emerge. The presence of petrifications suggested that insight might be gained regarding a longstanding debate over the origins and age of the Earth; the coming and going of birds, particularly swallows and martins, spoke to the broader

---

[71] Catherine Tatiana Dunlop, *Cartophilia: Maps and the Search for Identity in the French–German Borderland* (Chicago: University of Chicago Press, 2015), 139.

[72] Milton Anthony to Barton, June 22, 1809; VD-BSB APS.

[73] Thomas W. Barton to Barton, April 2, 1814; VD-BSB APS.

question of avian migration; and, of course, the presence of so many herbaceous plants help either add new species to existing catalogs, or provide useful information for those seeking to refine and improve the systems for botanical classification.[74] The enlightened project Barton's network had been created to advance, in other words, almost required that the Alabama territory be brought under the United States' control to allow its potential contributions to science to be discovered and promulgated.

The network that Benjamin Smith Barton established among scientists stretching from North America to Europe, South America, and Asia highlights both how much has changed and how much has not in the time from Franklin to Facebook. Certainly, the number of participants has expanded, to say nothing of the speed with which the information travels. At the same time, what was true of Barton's time about the potential effects and consequences of these networks remains true today. Their ability to link like-minded individuals around a set of common assumptions, largely independent from the direct control of a state-defined mechanism, provided a way to promote potentially constructively disruptive ideas and to facilitate the communication of useful information. At the same time, Barton's network shows us that the motives and the consequences of those disruptions and efficiencies are not always what they seem, whether they served to promote the scientific or imperial ambitions of the persons involved, or whether they limited information or marginalized other ways of knowing and thinking about the world. Perhaps most simply, Barton's network reminds us of the very basic principle that the ability to disseminate knowledge is power, and that any network's potential benefits or problems lie not in the technology behind it, but the humans who use it.

---

[74] The prominence of these subjects in contemporary discussion of natural history is described by Andrew J. Lewis, *A Democracy of Facts: Natural History in the Early Republic* (Philadelphia: University of Pennsylvania Press, 2011) and Winterer, *American Enlightenments*.

# 4

# Planting the Seeds of Empire: Botanical Gardens and Correspondence Networks in Antebellum America

## Alicia DeMaio

"I am busy as ever," Asa Gray wrote to George Engelmann in 1858, a common refrain in the hundreds of letters the two exchanged.[1] Answering letters occupied most of Gray's time—without letters filled with information, specimens, seeds, and live plant donations, he could not do any of his other work, including writing plant descriptions for publication. Correspondence networks had, for centuries, sustained botanical study.[2] They allowed botanists to exchange plants, seek funding for collecting ventures, and relay news and information to collectors in the field. They proved particularly essential for institutions such as botanical gardens, as they provided the primary mechanism by which such institutions acquired their collections. This chapter explores the hierarchies latent in antebellum American scientific correspondence networks. Like their British counterparts, American correspondence networks centered around a few people of power, shaped by either their wealth or expertise. The hierarchies in these American networks were more mutable, but power differentials were present nonetheless. Those who held positions of power maintained the network's structure through shared trust and certain standards used to establish authenticity. When eagerness for more specimens caused those in positions of power to ignore these authenticating standards, the network temporarily malfunctioned, though it never collapsed entirely.

Other histories of American science that have examined correspondence networks note patterns of deference during the Colonial period, in which all of the experts who resided across the Atlantic Ocean, replaced by a "democracy of facts" in the Early Republic, where "experts" did not really have much more knowledge than the farmers, enslaved people, and Indigenous peoples who gave them reports on natural occurrences.[3] However, the establishment of authority and expertise was still critical to early American science, especially during classificatory debates.[4] By the

---

[1] Gray to Engelmann, Aug 5, 1858, Missouri Botanic Garden.

[2] Correspondence networks also sustained business and social relationships; see Margaret Meredith, "Friendship and Knowledge: Correspondence and Communication in Northern Trans-Atlantic Natural History, 1780–1815," in *The Brokered World: Go Betweens and Global Intelligence*, ed. Simon Schaffer (Sagamore Beach, MA: Science History Publications, 2009), 151–92; Stefan Müller-Wille, "Nature as Marketplace: The Political Economy of Linnaean Botany," *History of Political Economy* 35, no. 5 (2003): 154–72; Müller-Wille, "Collection and Collation: Theory and Practice of Linnaean Botany," *Studies in History and Philosophy of Biological and Biomedical Sciences* 38, no. 3 (2007): 541–62.

[3] For the Colonial period, see Susan Scott Parrish, *American Curiosity: Cultures of Natural History in the Colonial British Atlantic* (Chapel Hill: University of North Carolina Press, 2006). Kariann Yokota argues that American deference for British naturalists spilled over into the Early Republic; see Kariann Akemi Yokota, *Unbecoming British: How Revolutionary America Became a Postcolonial Nation* (Oxford, UK: Oxford University Press, 2011), chapter 4. See also Andrew Lewis, *A Democracy of Facts: Natural History in the Early Republic* (Philadelphia: University of Pennsylvania Press, 2011).

[4] One such debate occurred in a New York courtroom in 1819, when *Maurice v. Judd* attempted to discern whether a whale was a fish and relied on expert knowledge as testimony, though the jury ultimately decided against the opinion of Samuel Latham Mitchill that a whale was not a fish. D. Graham Burnett, *Trying Leviathan: The Nineteenth-Century New York Court Case that Put the Whale on Trial and Challenged the Order of Nature* (Princeton, NJ: Princeton University Press, 2007).

mid-nineteenth century, American correspondence networks began resembling European ones because they valued contributors with expertise. *Experts* were defined as those who could claim high social status and education, access to knowledge, and the ability to provide that knowledge to others. Field work allowed collectors to establish credibility and enter the network, though they needed to be sponsored in order to embark on an expedition unless they were independently wealthy. Collectors, however, were not permitted to make conclusions about what they observed in the field; this was reserved for gentlemen experts.[5] These kinds of scientific networks inform the cycle of knowledge accumulation that Bruno Latour presents in his model of "centers of calculation," in which a voyager journeys into the periphery in order to bring back information that can be processed into knowledge in the geographical center of the empire, the metropole where the experts lived, surrounded by like-minded individuals and scientific institutions that held the collected information in the form of specimens.[6]

Attention to work on British collecting in the first half of the nineteenth century is instructive in understanding how the correspondence networks described in this essay functioned, and the hierarchies—but also the mutual dependence—embedded in them. Anne Secord's work on gift exchange between gentleman naturalists and artisans indicate how artisans were able to enter and sustain their place in the network by providing specimens to gentleman naturalists, but gentleman would always send more specimens in return, reminding artisans of their dependent status.[7] Jim Endersby, in his study of Joseph Hooker, argues that although Hooker resided in the geographic center of the empire, he lived far from the plants that he needed to do his work, requiring him to cultivate relationships with collectors. In these relationships, Hooker needed to maintain his status as the center of the correspondence network, the processor of knowledge rather than the collector of information, so he carefully managed the information he provided to his collectors. He gave just enough books and tools to allow them to collect useful specimens, but not enough that they could describe or name specimens, and he denied any requests to let them name their

---

[5] Jane Camerini, "Wallace in the Field," *Osiris* 11 (1995): 44–65; Stuart McCook, "'It May Be Truth but It Is Not Evidence: Paul du Chaillu and the Legitimation of Evidence in the Field Sciences," *Osiris* 11 (1996): 177–97; Jeremy Vetter, "Cowboys, Scientists and Fossils: The Field Site and Local Collaboration in the American West" *Isis* 99, no. 2 (June 2008): 273–303.

[6] Bruno Latour, *Science in Action: How to Follow Scientists and Engineers through Society* (Cambridge, MA: Harvard University Press, 1987), chapter 6.

[7] Anne Secord, "Corresponding Interests: Artisans and Gentlemen in Nineteenth Century Natural History" *The British Journal for the History of Science* 27, no. 4 (1994): 383–408.

collections.[8] Asa Gray, as this chapter demonstrates, used similar tactics in order to cultivate connections, facilitate his botanical work, and maintain his position as the center of American botanical correspondence networks.

American antebellum botanical networks had, roughly, three levels to their hierarchy. At the top sat the universal experts, such as John Torrey and Asa Gray, who worked to amass as much knowledge about American flora as they possibly could. Both Torrey and Gray were medical doctors, and they also held professorships: Torrey at Princeton and Gray at Harvard. Slightly subordinate to Gray and Hooker within this hierarchy are what I call the *specialists*—men like George Engelmann, Moses Ashley Curtis, and Charles Wilkins Short. These men had either regional or typological specialties—cacti and mycology for Engelmann and Curtis, respectively, and Kentucky flora for Short. These men were just as educated as Hooker and Gray—Engelmann and Short were both medical doctors. Engelmann, additionally, published his plant descriptions internationally, which allowed him to broaden his correspondence network overseas, especially with fellow German botanists.[9]

The one factor that distinguished the second tier of botanists from the first was how they regarded their own work—as hobby or profession. The distinction between "amateur" and "professional" scientists is a fraught one in the historiography of nineteenth-century science, and much ink has been spilled over the usefulness of these categories, as they are often used teleologically or to indicate superior or inferior scientific work.[10] It is useful, however, to look at the categories that members of network used to describe themselves, which often did fall into categories of those who could make a living from science, or otherwise devote full-time study to their chosen subjects, and those who could not, but were still members of

---

[8] Jim Endersby, *Imperial Nature: Joseph Hooker and the Practices of Victorian Science* (Chicago: University of Chicago Press, 2008).

[9] Oscar Soule, "Dr. George Engelmann: The First Man of Cacti and a Complete Scientist," *Annals of the Missouri Botanical Garden* 57, no. 2 (1970): 137.

[10] See, for example, George H. Daniels, "The Process of Professionalization in American Science: The Emergent Period, 1820–1860," *Isis* 58, no.2 (1967): 150–66; Sally Gregory Kohlstedt, *The Formation of the American Scientific Community: The American Association for the Advancement of Science, 1848–60* (Urbana: University of Illinois Press, 1976); Nathan Reingold, "Definitions and Speculations: The Professionalization of Science in America in the Nineteenth Century," in *The Pursuit of Knowledge in the early American Republic: American Scientific and Learned Societies from Colonial Times to the Civil War* (Baltimore, MD: Johns Hopkins University Press, 1976), 33–70. Paul Lucier has redefined professionalization to be a question of which scientists are paid for their work and which are not, since gentlemen experts often prided themselves on not being paid. Lucier, "The Professional and the Scientist in Nineteenth-Century America," *Isis* 100, no. 4 (December 2009): 699–732. Endersby argues that Hooker would not have referred to himself as a *professional* as the word was equated with the medical profession, and his expertise rested on disinterest and disdain for monetary gain from science. Gray and Torrey, on the other hand, constantly talked about their lack of money and how they wished they had more.

gentlemanly "professions."[11] Engelmann and Curtis, who fell into the latter category, certainly thought of their own inability to study botany full time as evidence of their subordinate status to those who could. Engelmann often lamented his inability to quit his medical practice, since he needed the income to support his family if he ever hoped to be financially "independent" and able to pursue botany full time.[12] When Charles Wilkins Short—who was independently wealthy—remarked on Engelmann's elevated place in the botanical hierarchy, Engelmann was quick to correct him: "You flatter me by far too much in associating me with Torrey and Gray and putting me with them at the head of present botanists."[13] After Short recategorized him with slightly lesser botanists, Engelmann replied that Short still did him "too great an honor ... though I confess that my desire would be to devote myself entirely to Botany," indicating that for Engelmann, full-time study allowed Gray the time and financial support to amass extensive knowledge of the field.[14] Similarly, Curtis wrote that he had "no ambition to be a Botanist" as he "meddled with flora only for recreation," distinguishing himself from others like Gray.[15] Scholar Elizabeth Keeney defines *professional* as someone motivated to change science and *amateur* as motivated by a desire to improve oneself through practicing science.[16] However, Curtis and Engelmann, who did not strictly make a living through science and who did not consider themselves "professional," still believed that they were contributing to scientific knowledge. Indeed, that is what drove them to spend so many hours outside of their chosen profession collecting, classifying, and corresponding with other botanists.

The third tier in the botanical hierarchy, reflected in the correspondence network, consisted of botanical collectors, some of whom held college degrees and some of whom did not. These men performed the physical labor of traveling into the field and collecting seeds and plants to prepare as botanical specimens. They would then sell the specimens to more eminent botanists like Short, earning a living—or attempting to—from their collecting. Engelmann, located in the gateway to the American West, often recruited these collectors, particularly for western expeditions. Sometimes, Torrey or Gray would find and recruit collectors using the correspondence

---

[11] Ruth Barton, "'Man of Science:' Language, Identity and Professionalization in the Mid-Victorian Scientific Community," *History of Science* 41, no. 1 (2003): 90–100, 107.

[12] Engelmann to Gray, Nov 14, 1853 and Dec 30, 1859, Gray Herbarium, Harvard University.

[13] Engelmann to Short, Nov 14, 1853, Charles Wilkins Short Papers, 1784–1879, 1951, Collection Number 662, Wilson Library, University of North Carolina, Chapel Hill.

[14] Engelmann to Short, June 12, 1855, Short Papers, UNC Wilson.

[15] Curtis to Gray, June 18, 1846, Gray Herbarium.

[16] Elizabeth Keeney, *The Botanizers: Amateur Scientists in Nineteenth-Century America* (Chapel Hill: University of North Carolina Press, 1992), 3–7

network. The network therefore both produced collectors and then expanded to incorporate them. Throughout the nineteenth century, the relationship between collector and expert shifted, as experts distanced themselves but still relied on collectors, rebranding them from "colleague" to "helper."[17] Collaboration still existed within the networks denoted by these hierarchies, especially between members of the network at the same level of expertise.[18] But collectors and experts like Gray also needed to collaborate, if they wanted their work to be done. However, these collaborations reveal inequalities of access to materials and information.

It should be noted that women also collected plants and entered the correspondence network in order to share their specimens, but they typically collected around the areas where they lived and did not make journeys under military escorts, unlike the male western collectors mentioned earlier. William Darlington, a former student of Benjamin Smith Barton and an enthusiastic supporter of the Cambridge Botanic Garden, was eager to introduce Gray to a "Lady Botanist" named Elizabeth Morris, who lived just outside of Philadelphia in Germantown. Darlington found Morris somewhat remarkable in her botanical success; he described her as possessing "zest & energy of intellect quite unusual in her Sex" that allowed her to have "a more extensive knowledge of Plants than any Female I have ever met with." Darlington was sure Gray "would be pleased to have such a correspondent as" Morris, and he was "satisfied you could engage her to make collections that would be useful to you."[19] Darlington wanted Gray to come to Philadelphia to meet Morris in person, but he was unable to make the trip. Darlington, ever persistent, begged him to at least write to her, "for the true lovers of Botany are too few, in our country, to be kept segregated. We cannot afford to be personal strangers to each other, when an acquaintance would contribute so much to our mutual advantage."[20] Darlington forwarded seeds from Morris and once again asked him to begin corresponding with her, in the form of a thank-you note. It is telling that Morris did not feel it was appropriate to address a letter to Gray on her own, without an introduction from Darlington and without Gray initiating their correspondence. This may have been due to her gender but also Gray's position at the top of the botanical hierarchy. Gray did eventually write to her, beginning a lengthy correspondence and exchanges of seeds and specimens.

[17] Keeney, *The Botanizers*, 30.
[18] Cairn Berkowitz, "The Endless Frontier: Joseph Leidy and the Collaborative Work of Natural History in Mid-Nineteenth Century America," in *Science Museums in Transition*, ed. Carin Berkowitz and Bernard Lightman (Pittsburgh: University of Pittsburgh Press, 2017), 218.
[19] William Darlington to Asa Gray, Sept 23, 1842, Asa Gray Correspondence, Harvard University Herbaria.
[20] Darlington to Gray, Oct 18, 1842, Asa Gray Correspondence, Harvard University Herbaria.

There were no strict rules regulating communication between the levels of the network's hierarchy—sometimes collectors or local botanists asked permission to write to more eminent botanists, but sometimes they just spontaneously wrote. Usually, Gray or Torrey would act as emissary if American botanists wanted to communicate with foreign ones. When Short asked Gray whether he would send a package to Hooker on his behalf, Gray replied, "it will always give me great pleasure to act as an intermediary in your correspondence with Hooker, or any other foreign botanist. Sir William is one of my nearest and dearest friends—now of almost 20 years standing."[21] Whether he was boasting or not, Gray's recognition of his role in the correspondence network as bridge between America and the rest of the world speaks to the power he held among American botanists.

Gray cultivated relationships of exchange with both collectors and naturalists-in-training, two distinct roles on the correspondence network hierarchy. With a collector, Gray established clear boundaries of what he expected—plants in exchange for money and/or acknowledgment of collection when he published the plant descriptions. Collectors were not permitted to name plants, and Gray—like Joseph Hooker—would often withhold books or tools in order to maintain this balance of power. Sometimes, collectors—especially those who had a long relationship with Gray—challenged their place in the hierarchy. Charles Wright was one such collector. Wright, who hailed from Connecticut and held a BA and MA from Yale, moved to Texas in order to teach school and began collecting plants.[22] He created a relationship with Gray after sending him some of his specimens, unsolicited; typically, Engelmann introduced western collectors to Gray, but it was not unusual in British correspondence networks for collectors to send specimens to a gentleman botanist in order to enter the network.[23] Gray was impressed with Wright's work and secured him a place on the Mexican Boundary Survey in 1850 and 1851. Wright, by then having made contacts in the nascent Smithsonian Institution, then joined the U.S. North Pacific Exploring Expedition.

In 1859, Wright left for his second of three trips to Cuba, at a time when Southern slaveholders were desperate to conquer the island in order

---

[21] Gray to Short, Feb 5, 1855, Short Papers, UNC Wilson.

[22] For more about Wright see Clinton P. Hartman, "Charles Wright: Botanizer of the Boundary Part 1: A Connecticut Yankee in Van Horne's Train," and "Charles Wright: Botanizer of the Boundary Part 2, At the Edge of the Power Struggle in the U.S. Boundary Commission" *Password: The Journal of the El Paso County Historical Society* 37, no. 2 (Summer 1992): 55–70 and 37, no. 4 (Winter 1992): 171–87; Samuel Wood Geiser, *Naturalists of the Frontier* (Dallas: University Press, 1948), 172–98.

[23] Soule, "Dr. George Engelmann," 138; Secord, "Corresponding Interests," 393.

to gain more slaveholding territory and supported multiple filibustering expeditions to this end. Wright often wrote of local suspicion surrounding his visit, which blocked his access to news and his ability to send letters. He wrote, in his typical blunt and colorful prose, "I don't know what has bewitched our correspondence unless it is filibusterism. I wish it was beheaded." Because American males created unease among local Cubans, Wright urged Gray to "*send* along Mrs. Gray" instead. Jane Gray was an active member of her husband's botanical network—she transcribed some of Gray's letters in her much more legible handwriting, she answered letters when Gray needed to run off to Europe to help her sick brother in Paris, and she corresponded with other botanists' wives in a kind of parallel version of the correspondence network. Wright, who had visited the Grays, also knew Jane Gray personally. He must have known that she had a penchant for tropical fruit, for he argued that if she visited Cuba, she could "have as many bananas as she wants & any other fruit to which *she takes.*"[24] Not surprising, Gray did not send his wife off to Cuba to stay with an unmarried man, though women did visit Cuba in the antebellum period to restore their health in a tropical climate, and Jane Gray suffered from digestive troubles all of her life.[25]

Wright, who was by now very familiar with Gray and who was often bored in Cuba during rainy weather, begged him to send books—so that he could learn enough about plants in order to name them, a violation of the correspondence network hierarchy. If Gray did not comply with his requests, Wright became angry. "You have a way peculiarly your own of dictating to me how I shall spend my time—stick to the woods *honey!* & let the books alone," he taunted. Gray had recommended that Wright study when he returned home, but Wright saw through this suggestion: He asked, "What'll be the use after they are all studied up named & distributed?"[26] Wright brought up naming plants again in another letter. "It is natural that I should desire occasionally to give a name to a plant where there seems a reasonable probability that it is undescribed," he argued. As if to indicate that he did not mean to fully usurp Gray's role, he added, "I don't like writing well enough to dabble much in description & then I haven't got a Latin dictionary in my '*cabeza*' to offer me the right word just when I want it. So I have sometimes just put down what I thought would be a good

---

[24] Wright to Gray, April 24, 1859, Gray Herbairum.
[25] Lisa Ann DeCesare, "Jane Lathrop Loring Gray (1821–1909) and the Archives of the Gray Herbarium," *Harvard Papers in Botany* 15, no. 2 (2010): 222. Sophia Peabody, future wife of Nathaniel Hawthorne, made such a visit with her sister over twenty years earlier in 1833. Megan Marshall, *The Peabody Sisters: Three Women Who Ignited American Romanticism* (Boston: Houghton Mifflin, 2005).
[26] Wright to Gray, May 14, 1859, Gray Herbarium.

name without any description."[27] Clearly, Wright wanted to stake his own
claim on the plants he collected, rather than wait for the botanist to name
a plant after himself, as was a typical way to honor the collector. Wright
also found other ways to challenge Gray's authority. After Gray insinuated
that Wright was "*dilly-dally*-ing," Wright reminded Gray that he was not
his military commander.[28] Gray, however, still retained social authority
over him, thanks to Gray's control over Wright's access to knowledge and
his financial patronage. Despite his denouncement of filibustering, Wright's
attempts to usurp the hierarchy in which he found himself resonated with
the antebellum ideology of martial manhood, in which violence and foreign
adventuring would lead to social mobility and future prosperity.[29]

Often absent in discussions of British correspondence networks, but
more present in discussions of American science during the Colonial period
and Early Republic, are the informants of the collectors, often Native
Americans or enslaved people.[30] In the antebellum period, collectors often
dismissed or attempted to erase the contributions of these populations,
indicative of the imperialist claiming ethos fueled by racist ideas of intellectual inferiority. Wright, who aligned himself with the White supremacist
Confederacy despite his Connecticut origins, often refused help from enslaved people offered to him in Cuba by his White hosts. When one of his
hosts "insisted on sending a negro with me to carry lunch &c.," on the
second day "I stole a march on him—went before he was up," unwilling
to accept the local knowledge of this man. By contrast, Wright welcomed
help from the "younger brother of the overseer" who brought him plants
"that I have not met with myself."[31]

Gray rarely mentioned Indigenous knowledge of plant uses in his
correspondence, except in the case of the saguaro cactus, a specimen of
which he grew in his greenhouse and wanted Hooker to grow at Kew. The
large cactus would serve as a symbol of the United States' imperial might
after its 1846 invasion of Mexico. In letters to Hooker, Gray notes how the
"Indians"—by which he means the Tohono O'odham—collected "the pulp

[27] Wright to Gray, June 30, 1859, Gray Herbarium. Gray ended up outsourcing descriptions of Wright's
Cuba plants to a German botanist, August Griesbach. See Richard A. Howard, *Charles Wright in Cuba,
1856–1867* (Alexandria, VA: Chadwyck-Healey, 1988).

[28] Wright to Gray, April 8, 1860, Gray Herbarium.

[29] Amy Greenberg observes that those who joined filibusters, such as William Walker, did so for a chance
at upward social mobility and economic gain. See *Manifest Manhood and the Antebellum American Empire*
(New York: Cambridge University Press, 2005), chapters 1, 2, and 4.

[30] Examples of such work include Cameron Strang, *Frontiers of Science: Imperialism and Natural Knowledge
in the Gulf South Borderlands, 1500–1800* (Chapel Hill, NC: University of North Carolina Press, 2018);
Judith Carney, "Out of Africa: Colonial Rice History in the Black Atlantic," in *Colonial Botany: Science,
Commerce, and Politics in the Early Modern World*, eds. Londa Schiebinger and Claudia Swan (Philadelphia:
University of Pennsylvania Press, 2005), 204–222; Parrish, *American Curiosity*, 2006.

[31] Wright to Gray, Nov 6, 1859, Gray Herbarium.

and seeds," which they made into "a sort of *jam*" and "use as an article of foods in the winter."[32] Gray related this piece of information as a novelty; official reports from the Mexican Boundary Survey, where Gray's specimen of the saguaro was collected, valued the agricultural skill of nations, such as the Tohono O'odham, as indicative of their future assimilation potential into "civilized" White society. Native Sonorans did not grow the right types of crops for the surveyors—meaning cotton, sugar, or other cash crops that can be supported by the slave economy.[33]

Gray's mentorship of William Henry Ravenel, cultivated through a relationship of exchange of specimens for knowledge and introductions, stands in direct contrast to Gray's relationship with Wright. Ravenel was a South Carolina planter who had wanted to become a doctor but was discouraged from this professional path by his father, who thought medicine would be too strenuous and gave his son a plantation to run instead. Ravenel's interest in agricultural experimentation lead him to botany, which lead him to a correspondence with other botanists, such as Curtis and Gray. Gray solicited specimens of South Carolina plants from him, explaining that, "it is ... very desirable that I should see as many plants from as many different localities as possible, and I have received much more material from the North & South & West of you, than from your state."[34] Gray's statement speaks to the mutually exclusive nature of the correspondence network, even among not-quite equals: Gray received the specimens he needed in order to complete his study of North American flora without having to travel across the country and collect the specimens himself, and in return, Ravenel received further education and an introduction to other botanists. In another letter, Gray connected him to New England botanists who studied Southern plants, and he promised to send Ravenel "collections of European plants ... if you care for foreign Botany."[35] Some American botanists, like Curtis, did not care for foreign botany, and immediately sent Gray any foreign plant specimens they received. But studying as many plants as possible would help refine the collector's eye, so accepting foreign plants from Gray was a valuable educational opportunity.[36]

---

[32] Gray to Hooker, 15 November 1853 and 28 March 1854, Director's Correspondence, Kew Gardens, DC/64/174 and DC/64/181.

[33] *Report of the United States and Mexican Boundary Survey Made under the Direction of the Secretary of the Interior, by William H. Emory.* Vol. 11, Part I (1857), 19, 117.

[34] Gray to Ravenel, August 24, 1846, University of North Carolina Department of Botany Historical Collections, 1810–1930s. Collection Number 02053, Wilson Library, University of North Carolina at Chapel Hill.

[35] Gray to Ravenel, December 23, 1846, UNC Wilson Library.

[36] On visual learning in natural history, see Lorraine Daston, "The Empire of Observation, 1600–1800," in *Histories of Scientific Observation*, eds. Daston and Elizabeth Lunbeck (Chicago: University of Chicago Press, 2011), 81–114; and Daniela Bleichmar, *Visible Empire: Botanical Expeditions & Visual Culture in the Hispanic Enlightenment* (Chicago: University of Chicago Press, 2012).

Ravenel, whose plants had been well received by Gray (he called the specimens "good," praise he did not use lightly), offered to collect samples of seeds and fruits for the book that Gray was writing, the *Illustrated Genera* of all US plants. Gray was delighted and promised to send "special demands," but also asked him to collect duplicates of anything he saw, as "they all come into play, & there is a great want of *fruit* in our collections."[37] Some of the seeds Gray actually planted in the Cambridge Botanic Garden after he examined them for the *Genera*, such as those from *Lupinus villosus*, or lady lupine, a lovely conical plant with purple blossoms. Gray was especially interest in specimens from the *Magnolia grandiflora*, or Southern magnolia, with its beautiful, large white blossoms. Gray sent to Ravenel both his desiderata as well as specific instructions for how to send what he wanted: "I hope you will be able to send me cores of Magnolia grandiflora—the only one I have not here. Can you not put a bud or two, and a just opened flower into a jar or wide-mouthed bottle of spirit.—also a core when full grown, & stop with large cork or plug, cover with resin or wax—& send me." Gray explained why exactly he wanted the flower and the core: his assistant and illustrator, Isaac Sprague, "is to figure anew this season all our Magnolias & I want this also direct from nature. I have a tree in green-house—6 feet high—but it does not flower."[38] This statement speaks to both the importance and the limits of the botanical garden for scientific study. Gray generally preferred to study plants in a living, growing state, which is why he tried to grow the lady lupine. But plants from elsewhere did not always grow in Cambridge, and when they did, they did not always behave as they would in their natural habitat. Dried plants, however, would occasionally lose their color or shape, which is why Gray specifically wanted a fresh flower preserved in a jar. When Ravenel did send the specimen, Gray signed his letter in response "Your obliged friend," as opposed to the more formal "Yours very truly and respectfully" of previous letters, showing his appreciation.

Despite this profession of friendship, Ravenel still showed deference to Gray, respecting his place in the botanical network. When Ravenel sent him a botanical question, he would often preface his question with a phrase such as, "I trust I am not troubling you unnecessarily."[39] Gray replied, "Do not concern yourself about *troubling* me. I am always very glad to hear from you—anxious to share in your botanical novelties, & ready to

[37] Gray to Ravenel, April 9, 1847, UNC Wilson.
[38] Gray to Ravenel, May 23, 1849, UNC Wilson.
[39] Ravenel to Gray, April 2, 1852, Gray Herbarium.

answer your queries as far as I can, in my hasty way."[40] Gray was always very clear about how busy he was, and yet he took the time to answer letters from a range of botanists. These claims of busy-ness served to remind the other botanists of Gray's prestige—making them grateful for his help and respectful of his position of authority.

Charles Wilkins Short, despite not being as expert as Gray, had a special place in the botanical correspondence network because he possessed something few other botanists did—money. Short had come into money after the death of his uncle, William Short, a close friend of Thomas Jefferson. With this inheritance, he expressed interest in supporting botanical causes. This was exciting news in the correspondence network, as there was a running joke in botanical circles about the lack of money available to support collecting expeditions or publications. Short himself made a joke about William Sullivant, a botanist from Ohio, being a "rara avis among naturalists" because he was a "man of fortune."[41] John Torrey immediately began providing Short with causes he could support, including the Mexican Boundary Survey.

When collectors approached Short directly asking for money, Short always conferred with Torrey about their legitimacy, indicating how those at the top of the network's hierarchy trusted each other to vouch for the authenticity of those who joined—in lieu of formal letters of introduction.[42] When Samuel Botsford Buckley, a botanical collector who frequently made excursions into the southeastern and western states, asked Short to financially support a planned journey to Utah and California, Torrey explained that he had a longstanding correspondence with Buckley, lasting "eight or ten years." He cited his vast collection experience and his "zeal" for botany. Torrey explained, "I believe him to be an honest man, as he is certainly a zealous collector." Although Torrey acknowledged that his specimens needed work, he thought Buckley's preservation skills would improve with more means at his disposal. Overall, Torrey thought Buckley was "quite trustworthy, has had much experience as a collector, & is pretty well acquainted with North American Botany," his second emphasis on Buckley's good character.[43] George Engelmann, however, offered another and more succinct comment on Buckley's collection potential: "I wish we had a trustworthy botanist, to collect about the Salt Lake; I would pay a

---

[40] Ravenel to Gray, April 23, 1852, UNC Wilson.
[41] Short to Gray, Dec 26, 1842, Gray Herbarium.
[42] Letters of introduction were particularly important in European correspondence networks. Secord, "Corresponding Interests," 384.
[43] Torrey to Short, Sept 21, 1850, UNC Wilson.

small share with pleasure; but Buckley makes very poor specimens."[44]
Ultimately, Short decided not to support Buckley, and true to his word,
Buckley journeyed west anyway; Torrey reassured Short that "he is not
offended with *you*."[45] When it functioned correctly, the botanical network
could be used to vet others and facilitate the work of botany through the
exchange of money as well as plants.

Torrey, in his eagerness to obtain new specimens, often was too quick
to trust collectors he did not know, endangering the correspondence network
in the process. In 1851, Torrey wrote to Short with yet another charity
case—a Swedish botanist named F. H. Lundgren. Torrey begged Short to
buy some specimens from Lundgren, who apparently trained with Elias
Fries in Uppsala, but who came to the United States with little money.
After sending $30 to buy Lundgren some new clothes so he could find
employment, Short somehow decided to fund an expedition to the mountains
of North Carolina in order to find the elusive *Shortia galacifolia*, the
flowering perennial named after him by Asa Gray. Andre Micheaux initially
collected and named the flower in the 1790s, but no one had seen it since.
After Short sent $100, Torrey reported that Lundgren had "a very good
outfit, & will, I have no doubt, do well."[46] By May of 1852, he had already
departed from Charleston. Lundgren wrote a letter to Short thanking him,
in broken English, for "the liberality and love to the lovely Science,"
and promised that he would do his best "in fulfilling the purpose of
my journey."[47]

In total, Short sent $400 in checks to fund Lundgren and his journey
over the summer of 1852. In September of 1852, Torrey remarked that he
had not heard from Lundgren, and assumed he "must be on his way
home."[48] By November, there were still no letters from Lundgren—and
Torrey began to panic. Was he sick? Did he die? They knew, from a friend
in Asheville, North Carolina, that Lundgren had been seen there, and he
had apparently left "a large bundle of plants" with said friend. Torrey
began to fear the worst—that they had been scammed, and that Lundgren
had left the United States, perhaps for Havana, Cuba, with Short's money.
Finally, in December of 1852, Torrey faced his fears: "We have all been
deceived, I fear, by the Swedish botanist!" he declared to Short. Torrey
tried to soften the blow by reporting that their contact in Asheville, Dr.

---

[44] Engelmann to Short, Nov 28, 1850, UNC Wilson.
[45] Torrey to Short, Aug 7, 1851, UNC Wilson.
[46] Torrey to Short, May 1, 1852, UNC Wilson.
[47] Lundgren to Torrey, May 8, 1852, UNC Wilson.
[48] Torrey to Short, Sept 18, 1852, UNC Wilson.

Hardy, did not want to call Lundgren "a bad man," and said he "was very favorably impressed by his intercourse with him. He would have trusted him with all his property."[49] Asa Gray, in a satisfactory "I told you so" moment, called Lundgren a "Swedish scamp" and added, "I never quite liked that he should come to this country direct from Sweden & not bring letters from Fries."[50] This comment illustrates that Torrey ignored the mechanisms of the correspondence network—the use of letters of introduction to vouch for a person's credentials. While Torrey was embarrassed, at least he did not lose $400.

Scientific networks in the antebellum period were concerned with fraud, often in terms of fraudulent knowledge—presentation of information depended very much on the social status of the person presenting, and to what extent they obeyed the norms of the network.[51] For the network to function at all, however, its members needed to be able to trust the opinions of others, especially those who ranked higher than them. All networks built on trust, whether business or scientific, opened themselves up to deceit, as Herman Melville brilliantly satirized in his novel *The Confidence-Man*, in which a series of passengers are conned on a steamboat up the Mississippi River.[52] Thankfully, this kind of scam was a rare occurrence, as correspondence networks could also be mobilized in order to quickly spread the word about fraudsters.[53] Short continued to support collectors he trusted; another one of Torrey's charity cases, a German botanist, ended up opening a pharmacy and thanked Short for his monetary aid.

Short also provided seeds to nurseries, institutions that forged connections with botanical circles and participated in correspondence networks through the antebellum period, using them to acquire the latest stock for their businesses. Robert Buist, a nurseryman in Philadelphia, thanked Short for the "fine parcel of seeds of *Virgilia*," saying of the gift, "I do not know anything ... that has gratified me more for many years." Stock of the *Virgilia* was scarce in the Philadelphia region, as the tree on the old Bartram garden grounds—now owned by industrialist Andrew Eastwick—flowered "every Season but rarely produces perfect seeds." In the spirit of friendship and exchange that structured the nineteenth-century horticul-

---

[49] Torrey to Short, Dec 23, 1852, UNC Wilson.

[50] Gray to Short, July 1, 1853, UNC Wilson.

[51] Lukas Rieppel, "Hoaxes, Humbugs, and Frauds: Distinguishing Truth from Untruth in Early America" *Journal of Engineering Research* 38, no. 3 (Fall 2018): 501–04. On trust in other communication networks, see Lindsay O'Neill, *The Opened Letter: Networking in the Early Modern British World* (Philadelphia: University of Pennsylvania Press, 2015).

[52] For more on financial fraud in the period, see Steven Mihm, *A Nation of Counterfeiters: Capitalists, Con Men, and the Making of the United States* (Cambridge, MA: Harvard University Press, 2007).

[53] Secord, "Corresponding Interests," 392.

tural trade, Buist promised to send in "return of any article in my way with sincere thanks for your kind remembrance."[54]

The following year, Thomas Meehan—who began his own nursery business after working for Eastwick at the former Bartram garden—also asked Short for seeds of the *Virgilia latea*. Clearly word spread that Short provided Buist with the seeds. Tellingly, Meehan introduced himself as a friend of William Darlington, a well-established botanist.[55] Meehan used his connection with one node of the botanical network to establish another connection within it. He sent his catalog, probably both for advertising and to prove that he actually had a nursery business and was thus a legitimate correspondent. He followed up this request asking whether Short knew someone who would collect seeds of *Cladrastris tinctoria*, a beautiful flowering tree that was assuredly in high demand in the nursery business. When Short sent more, Meehan reiterated his request for a collector, but also thanked him for the "fine 'well filled' seed" that would "no doubt ... greatly increase our stock."[56]

The botanical correspondence network in the antebellum United States was constructed by hierarchical relationships between the botanists who participated in it. The top tier consisted of experts in national flora such as Asa Gray and John Torrey, followed by medical doctors who gained renown for studying specific botanical regions or types such as George Engelmann and Charles Wilkins Short. Below them sat collectors who gathered plants and seeds without extensive study beyond what was necessary to identify plants of interest in the field. The boundaries of this hierarchy were, to a small extent, nebulous—there were no barriers of correspondence, and exchanges of information happened between levels of the hierarchy, but the power to publish, and especially to name plants, rested at the top. These networks facilitated the practice of botany by funding explorations and circulating seeds and live plants that would grow into the collections of a botanical garden, such as the Cambridge Botanic Garden, which Asa Gray superintended at Harvard. All members of the hierarchy, from Charles Short to Charles Wright, contributed seeds to Gray's garden.

---

[54] Robert Buist to Charles Wilkins Short, Dec 22, 1853, Charles Wilkins Short Papers, 1784–1879, 1951, Collection Number 662, Wilson Library, University of North Carolina at Chapel Hill. Marina Moscowitz, "Calendars and Clocks: Cycles of Horticultural Commerce in Nineteenth-Century America" in *Time, Consumption, and Everyday Life: Practice, Materiality and Culture* eds. Elizabeth Shove, Frank Trentmann, and Richard Wilk (Oxford: Berg, 2009), 116.

[55] Thomas Meehan to Charles Wilkins Short, June 24, 1854, Short Papers, Collection 662, Wilson Library, UNC Chapel Hill.

[56] Meehan to Short, Nov 30, 1854 and Dec 27, 1854, Short Papers, Collection 662, Wilson Library, UNC Chapel Hill.

# 5

# Science, Skepticism, and Societies: The Politics of Knowledge Creation in the Early Republic

## George D. Oberle III

T he United States was home to three major learned societies by 1800. The oldest, first established in 1743, the American Philosophical Society (APS), based in Philadelphia, sought to "Promote Useful Knowledge Among the British Plantations in America." Its leaders hoped the APS would serve as a single hub to promote scientific learning. Despite this expectation, and short periods of inconsistent activity, the membership became dormant until the Imperial Crisis reinvigorated a desire for American scientific independence from the British.[1] It is no coincidence that upon the approval of the Massachusetts Constitution of 1780, the people of that state empowered the government to promote a host of knowledge institutions, including learned societies, and that the Massachusetts legislature charted the Cambridge-based American Academy of Arts and Sciences (AAAS) on 4 May 1780. Connecticut's legislature finally approved the charter after several years of failed attempts, which began in 1781, for the Connecticut Academy of Arts and Sciences (CAAS), the third universal learned society, established in 1799.[2] The republic was expanding its institutional societies to promote knowledge. Why were they regionally based? Would these institutions with similar missions promote an apolitical scientific agenda? Would they promote similar types of knowledge?

These institutions had similar missions, but their membership drove both their political motivations and their research agendas. Membership in all three remained exclusive to the White elites, yet there was a changing understanding of who should create knowledge and who should have access to that knowledge. The emergence and expansion of these new institutions and their transformation provided evidence of an ongoing information revolution, which was analogous to the political and market revolutions of the eighteenth and nineteenth centuries. The idea of an information revolution

---

[1] "A proposal for Promoting Useful Knowledge among the British Plantations in America," *The Pennsylvania Gazette*, January 28, 1768, 2. The secondary literature on the history of the establishment and purpose of the American Philosophical Society is vast. Edward Carlos Carter, *"One Grand Pursuit": A Brief History of the American Philosophical Society's First 250 Years, 1743–1993* (Philadelphia: American Philosophical Society, 1993); Brooke Hindle, *The Pursuit of Science in Revolutionary America, 1735–1789* (Chapel Hill: Published for the Institute of Early American History and Culture, Williamsburg, VA by University of North Carolina Press, 1956). See Hindle's dissertation and other early histories of the APS. The APS never lived up to its goal of being the single learned society for the British Colonies in North America.

[2] Mary Ellen Ellsworth, *A History of the Connecticut Academy of Arts and Sciences, 1799–1999* (New Haven, CT: Connecticut Academy of Arts and Sciences, 1999). It is important to note that there were other societies that were established for the promotion of knowledge but they were more specialized. In fact, Stiles and Reverend Nathan Strong corresponded in 1780, the same year as the founding of the AAAS, about forming a society associated with Yale. The following year he presented a draft charter to the legislature. The charter failed to pass both houses of the legislature, which led to several attempts to establish the institution. Finally, in 1799, four years after the death of Stiles, it was passed. The group was and remains strongly associated with Yale.

remains understudied as compared to the Enlightenment.[3] It resulted from the expanding demand for information, which was fulfilled by an ever-expanding supply of information. The consumers for this new information included the traditional elites, government officials, and military officers, as well as a growing number of professionals with a need for specialized information. They, in turn, created new information and fed into an expanding information cycle. Further, there was a dramatic growth in population, production, and trade. In addition, as Daniel Headrick points out, "This period saw a substantial increase in the number of educated people, ... who judged one another by their conversations, their wit, their knowledge of the world and the latest news."[4] Information possessed a social and cultural power for those who had access to the right information. A learned society became an essential educational extension for those who finished their college studies and entered professional life. The wide range of new information systems, including newly established learned societies, helped to lay a foundation for a diverse information environment.

Scholars have detected ideological differences among the types of scholarship of the members of early American learned societies, which manifested in the political arena. Members of the APS and the AAAS showed this in their political actions. Thomas Jefferson was elected to the presidency of the APS after narrowly losing his 1796 bid for the presidency of the United States to John Adams, who served as the president of the AAAS. The ideological divide between Federalists and Republicans appeared most vivid when they debated what aspect of science ought to be considered an appropriate course of study, and, perhaps even more important, who should act as a gatekeeper to the authority that knowledge provided. Linda Kerber's *Federalists in Dissent* shows that the Federalists feared "that an ordered world was disintegrating; that a cherished civilization was imperiled" by the Jeffersonians because of their rejection of classical learning and focus on the wrong type of scientific endeavors.[5] When Jefferson received notification of his election as president of the APS, he replied that the "suffrage of a body which comprehends whatever the American world has of distinction in philosophy and science, in general,

---

[3] See Daniel R. Headrick, *When Information Came of Age: Technologies of Knowledge in the Age of Reason and Revolution, 1700–1850* (Oxford: Oxford University Press, 2000); George D. Oberle, "Institutionalizing the Information Revolution: Debates over Knowledge Institutions in the Early American Republic" (PhD diss., George Mason University, 2016).

[4] Daniel R. Headrick, *When Information Came of Age: Technologies of Knowledge in the Age of Reason and Revolution, 1700–1850* (Oxford: Oxford University Press, 2000), 10.

[5] Linda K. Kerber, *Federalists in Dissent: Imagery and Ideology in Jeffersonian America* (Ithaca, NY: Cornell University Press, 1970), 174.

is the most flattering incident of my life." Jefferson longed to be taken seriously as a man of learning and science, and he assured his colleagues that despite feeling "no qualification for this distinguished post but a sincere zeal for all the objects of our institution," and that he would work diligently to "see knowlege so disseminated through the mass of mankind that it may at length reach even the extremes of society, beggars and kings."[6]

Jefferson believed that extensive access to knowledge, specifically scientific knowledge, offered opportunities for progress to its citizenry and that societies, such as the APS, offered a means to promote scientific knowledge throughout the new republic. The Federalists, however, saw evidence of democratic science as both a threat to their political standing and the advancement of knowledge. In a series of pieces by Josiah Quincy, in the Federalist organ the *Port-Folio*, published by Joseph Dennie, the APS fell under scrutiny for not including a number of "men, full as distin-guished as any member, of the APS and that the "members of that society were not famous for A PARTICULAR AND ACCURATE KNOWLEDGE OF ANY THING, but Were distinguished only by a *general acquaintance* with philosophy."[7]

This apparent disagreement intensified in the first decade of the nineteenth century. The so-called "articulate Federalists" were fearful of the naivety of the Jeffersonians, who were providing access to specialized knowledge to those who were not properly prepared to consume this informa-tion. Worse still, the Jeffersonians allowed these unqualified men to hold the authority to create and disseminate knowledge. They saw a world filled with "roving natural historians, aimlessly searching for curiosities—bones, salt mountains, weird forms of animal life."[8] Federalists viewed such in-dividuals as charlatans masquerading as scholars. This battleground oc-curred in the pages of newspapers, and the Federalists used satirical accounts to attack the validity of natural history and other new scientific studies that were coming from the APS. The proliferation of these new scientific studies came as more explorations of the west occurred. All of which were sensational and centered around the people who associated with the APS. These "curiosities" found in the new territories of the

---

[6] Letter from Thomas Jefferson to American Philosophical Society, 28 January 1797. *The Papers of Thomas Jefferson Digital Edition*, eds. James P. McClure and J. Jefferson Looney (Charlottesville: University of Virginia Press, Rotunda, 2008–2019). Accessed January 19, 2019, http://rotunda.upress.virginia.edu/founders/TSJN-01-29-02-0218.

[7] Josiah Quincy, "CLIMENOLE," *Port-Folio* 4, no. 37 (September 15, 1804): 291. Also see Linda K. Kerber, *Federalists in Dissent; Imagery and Ideology in Jeffersonian America* (Ithaca, NY: Cornell University Press, 1970), 67–94.

[8] Kerber, *Federalists in Dissent*, 134.

Louisiana Purchase became significant opportunities to criticize both the object of appropriate study and to attack the critical issues of constitutional authority, the expansion of slavery, and even secession of northern states.[9] Historian John C. Greene counters the claims of these scholars, suggesting that Federalists did not have a problem with the study of Natural History. Instead, however, they demanded a "balance between science and the classics."[10] In fact, Greene argues that both groups were eager to receive work from all types of scientific studies.

The biggest challenge for scholars is accounting for the vast scope of published content coming from these institutions. In addition, the sheer number of members belonging to the these groups makes it difficult to determine the types of research interests the members held and whether they held shared memberships in different societies. This complexity may help explain the contradictions in the existing historiography. Digital scholarship offers a solution to dealing with the massive scale of the source pool. It can easily allow scholars to add in new relevant groups, such as those in the CAAS and the select group of men associated with Yale College led by Timothy Dwight, who committed themselves to attack Jefferson and those who publicly aligned with him.[11]

This analysis includes examination of all the publications of these three societies through 1810.[12] The APS produced six volumes containing about 487 essays. The other two societies produced far less. The AAAS produced four volumes with 158 essays and the CAAS only a single volume of 17 essays. Linda Kerber's important work, *Federalists in Dissent*, published in 1970, compared two volumes, one by the APS and one by the AAS, published between 1799 through 1804. She found that these years served as critical evidence of the conflict between the Federalists and the Jeffersonian Republicans. An apparent limit of this study is that there are only a few years of evidence available, which ignores potential evidence of the intellectual divisions that occurred. It also ignores the work being produced by the newly established CAAS, whose papers were being delivered during this crucial period but not published in their *Memoirs of the*

[9] David Dzurec, "Of Salt Mountains, Prairie Dogs, and Horned Frogs: The Louisiana Purchase and the Evolution of Federalist Satire 1803–1812," *Journal of the Early Republic* 35, no. 1 (February 19, 2015): 79–108.

[10] John C Greene, *American Science in the Age of Jefferson* (Ames: Iowa State University Press, 1984), 414–15.

[11] Christopher Grasso, *A Speaking Aristocracy: Transforming Public Discourse in Eighteenth-Century Connecticut* (Chapel Hill: Published for the Omohundro Institute of Early American History and Culture, Williamsburg, Virginia, by the University of North Carolina Press, 1999), 380–81.

[12] The APS did not publish a new volume until 1818. The AAAS does not publish a new volume until 1833. The CAAS did in 1910.

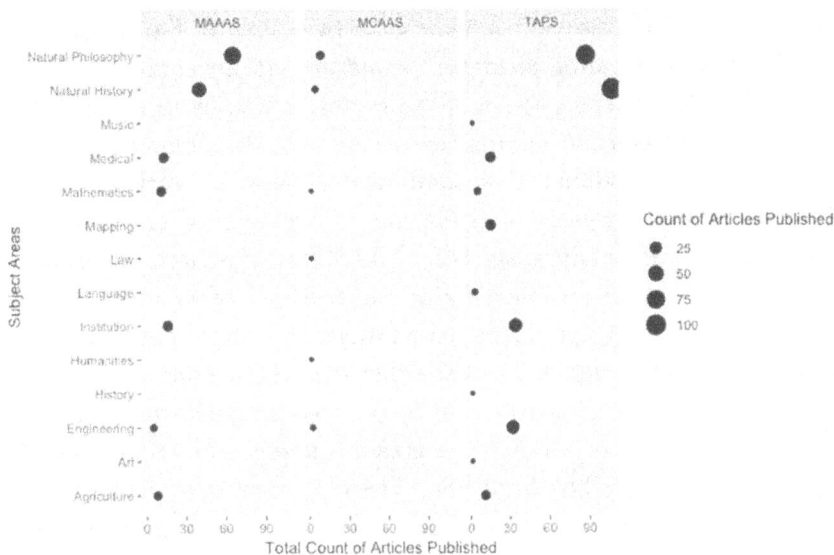

Figure 5.1 Total number of articles published by the early American learned societies, 1769–1810, faceted by subject.

*MAAAS, Memoirs of the American Academy of Arts and Sciences; MCAAS, Memoirs of the Connecticut Academy of Arts and Sciences; TAPS, Transactions of the American Philosophical Society.*

*Connecticut Academy of Arts and Sciences* until 1810. In performing this expanded analysis, we can test Kerber's findings, which asserted that the AAAS stressed natural philosophy, whereas the APS focused on the descriptive sciences of natural history.[13] Natural history emphasized life sciences and what we would now call *anthropological* and *paleontological studies.* Natural philosophy emphasized astronomical studies and chemistry.[14]

The graph in Figure 5.1 indicates the total number of articles published in the journals of the three societies by subject—the more articles published on a topic, the larger the representative bubble. There are a few points

---

[13] Linda K. Kerber, *Federalists in Dissent; Imagery and Ideology in Jeffersonian America* (Ithaca, NY: Cornell University Press, 1970), 76; John C. Greene, *American Science in the Age of Jefferson* (Ames: Iowa State University Press, 1984), 415. It is also important to note that there is no systematic study of the literature produced by subject; therefore there are other opportunities for study that connect to this historiographical tradition.

[14] This section uses a collection of data gathered from the specific publications of the three societies classified by subject. Using a package in Rstudio called *ggplot* to produce graphs allows for visualization and a careful analysis of the publication records of these societies. The implications of this method allow an expanded analysis to include other groups beyond the three learned societies discussed in this essay.

that stand out from this data. First, the APS published significantly more content than the other societies, more than twice that of the AAAS, most of which focused on natural history. This finding partially confirms Kerber's findings that the scholars at the APS privileged studies in the emerging fields of life sciences and anthropology. However, it is interesting to note that the APS also published a significant amount of work on natural philosophy/natural sciences. This domain, in Kerber's work, was one that was privileged by the Federalists in the AAAS. It is true that upon examination of the articles in the *Memoirs of the Academy of Arts and Sciences (MAAS)*, that the AAAS produced more articles on natural philosophy than natural history, However, we can see that the APS's *Transactions of the American Philosophical Society (TAPS)* published slightly more than the *MAAS*. The APS simply published a greater number of total articles, 487 essays compared to 158 by the AAAS. Therefore Kerber's finding that the AAAS privileged Natural Science/Philosophy is not validated by the data.

One way to control for the relative size of the journals is to look at the percentage of publications that each journal dedicated to a subject. Of 487 total articles, *TAPS* published 107 essays on natural history and 87 on natural philosophy, meaning that 22% of the articles concerned natural history and 18% involved natural philosophy. *MAAAS*, on the other hand, published 40 articles on natural history and 65 on natural philosophy out of 158 (Figure 5.2). This analysis shows the AAAS dedicated 25% of its journal content to natural history and 41% to natural philosophy. This result suggests that the AAAS is not necessarily against publishing on the subject of natural history and may indicate that perhaps there is less diversity of content published in *MAAS* than anticipated.

When adding the CAAS's publication, *Memoirs of the Connecticut Academy of Arts and Sciences*, to our analysis, what is clear is that natural history and natural philosophy are both incredibly important to all three groups and that these societies are less interested in most other forms of knowledge. What we can see from the figure is that what we now think of as scientific discourse is the essential form of knowledge represented in all of these publications. Figure 5.3 shows the total number of articles focused on natural history and natural philosophy in the three journals. The graph clearly shows the significant number of essays produced in the journals. Also noteworthy is that in its single volume of seventeen published articles, the *MCAAS* published eight natural philosophy articles, 47%, and four articles, 23%, on natural history. Together these twelve articles comprised 70% of the content. These findings seem to confirm Greene's argument that all the learned societies were interested in promoting scien-

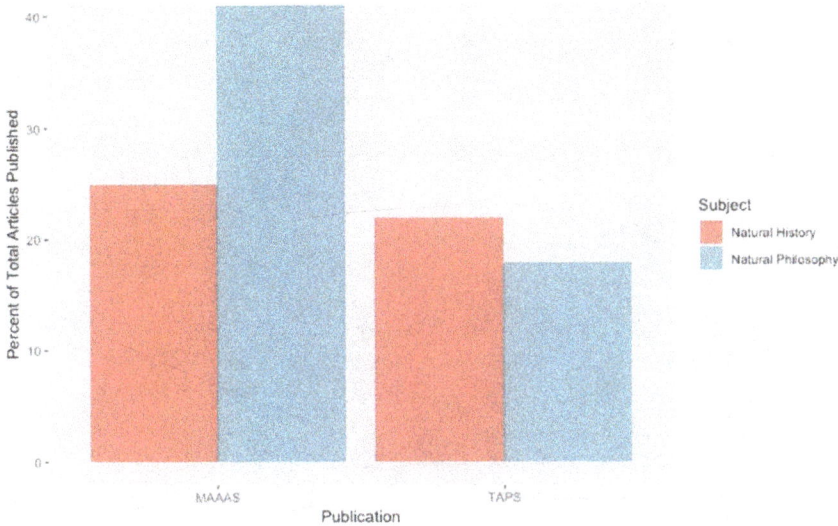

Figure 5.2 Percentage of articles published by *Transactions of the American Philosophical Society (TAPS)* and the *Memoirs of the American Academy of Arts and Sciences (MAAAS)* concerning natural history and natural philosophy.

tific scholarship. Still, there were distinct differences among the groups, which generally confirms Kerber's work.

Using network analysis and the membership lists of these groups, we can explore this issue differently and can start to see some distinctions. These sources are important and underutilized as evidence due to the difficulty in using them. For example, often the way members' names were written varied, thus making it difficult to construct a valid set of data to conduct an analytical survey . The secretaries of the organizations typically kept membership lists. The rules of membership varied, but typically required election to the society by current members of the body; these members paid an annual membership fee to remain in good standing. Membership dues paid for maintenance of buildings, to publish the journals, and to build and maintain the societies' collections of books, periodicals, manuscripts, and artifacts. These membership lists were often published. However, these published lists typically only indicated that a member was elected and not when a member ceased to pay his dues. These details can typically only be discerned from the original manuscripts maintained by the organizations' secretaries. As a result, knowing the actual membership

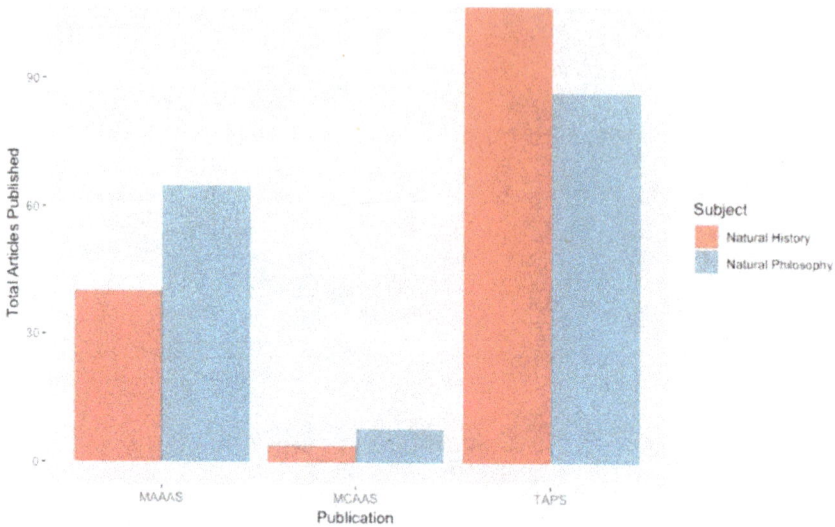

Figure 5.3 Total number of articles about natural history or natural philosophy published by all three society publications, including the *Transactions of the American Philosophical Society (TAPS)*, *Memoirs of the American Academy of Arts and Sciences (MAAAS)*, and the *Memoirs of the Connecticut Academy of Arts and Sciences (MCAAS)*.

list of a given society is difficult because these records were inconsistently managed. As the rolls grew, it became more difficult for the scholar to analyze membership data without the assistance of a computer.

Securing the most definitive list of members is more difficult than one may realize. The APS website has an online database of elected members, which contains a reliable list of elected members starting in 1768 available by year.[15] The database includes name, year elected, residency membership, living status, and death date.[16] The problem is that the members elected before the APS was reconstituted in 1768 do not show

---

[15] American Philosophical Society, "American Philosophical Society Member History," accessed 3/5/2019, https://search.amphilsoc.org//memhist/search.

[16] These fields are fairly self-explanatory; however, the residency field required exploration of the rules of the Society of the time. An elected member could hold one of two statuses based on the time of the election. A *resident* was a resident of the area that became the United States, whereas *international status* was bestowed upon anyone who lived outside that region. Over time, the membership rolls became more complicated; however, for this study these are the relevant core fields. Because there was no digitized list of resources or any way to export the list, a new spreadsheet needed to be created using the member's name, year elected, and residency. Scholars must make choices about whether to include all the data available or only parts of it and then explain these choices. In this case, the living status and the death date seemed irrelevant to the research questions being asked and would add extra time to prepare the data, therefore, it was not included.

up in the list by year, thereby requiring specific searches by name in the database. Using a published list of members from 1865 provided a means to access those names and a way to verify the data that I created.[17] The AAAS maintains a membership list online in its *Book of Members*.[18] Members are listed by name and year elected. The CAAS included a list of its members in its first publication, published in 1810 and titled, *Memoirs of the Connecticut Academy of Arts and Sciences*.[19] There are very few commonalities in the data that allow for comparison. Generally, these are the names of members and that they belong to a learned society.[20]

Examining the number of shared memberships can allow us to see whether the members of these scientific societies resisted electing their political opponents into their own learned society. There are fifty-one shared relationships between the APS and the AAAS, including Thomas Jefferson and John Adams. As seen in Figure 5.4, there are far fewer shared memberships between the other groups. This overlap likely occurred because during the Colonial period and the Confederation era, the leading political men of the era were often also viewed as leading scientists and thinkers. Few saw a distinction between politics and science. By the time the CAAS was established in 1800, a surprising change occurred. This change is evident from the lack of election of political opponents to the CAAS. There are few shared memberships between the CAAS and the other two organizations. There are only three shared connections among the three organizations. The CAAS and the AAAS had five connections although they only shared a single membership with the APS (Figures 5.5 and 5.6).[21]

---

[17] American Philosophical Society, *List of Members of the American Philosophical Society, Held at Philadelphia, for Promoting Useful Knowledge* ... (Philadelphia: The American Philosophical Society, 1865), https://catalog.hathitrust.org/Record/007077858. In two previous projects that I worked on using this data it turns out I was missing the names of 250 members who were part of the "original society." It is possible that this data exists elsewhere but this issue speaks to the ongoing challenges of establishing authoritative lists of data and sources, which Roy Rosenzweig identified in his 2003 article: Roy Rosenzweig, "Scarcity or Abundance? Preserving the Past in a Digital Era," *American Historical Review* 108, no. 3 (2003): 735–62.

[18] American Academy of Arts and Sciences, *Book of Members 1780–Present*, https://www.amacad.org/members/book 3/13/2019. In addition, the AAAS has a database that serves as a membership directory, which can be consulted in a variety of ways. See https://www.amacad.org/directory.

[19] Connecticut Academy of Arts and Sciences, "Table of Contents," *Memoirs of the Connecticut Academy of Arts and Sciences*, 1, no. 1 (1810): viii.

[20] In the case of the APS and the AAAS the year of election to the society is listed.

[21] Although there are two shared relationships between the CAAS and the APS, one was a false hit. John Allen turns out to be a false connection. John Allen of the American Philosophical Society was a founding member who died in 1778. It is unclear who the John Allen listed in the Connecticut Academy of Arts and Sciences is, yet it seems clear he is a different person based on the early death of the APS member of this name. Whitfield Jenks Bell, *Patriot-Improvers: Biographical Sketches of Members of the American Philosophical Society. Vol. 3 1767–1768* (Philadelphia: American Philosophical Society, Press, 2010). The only shared relationship between the CAAS and the APS involved a man named Charles Chauncey, who was likely the lawyer and son of the elder Charles Chauncey, a famous New England minister. The younger Chauncey was elected to the APS in 1813.

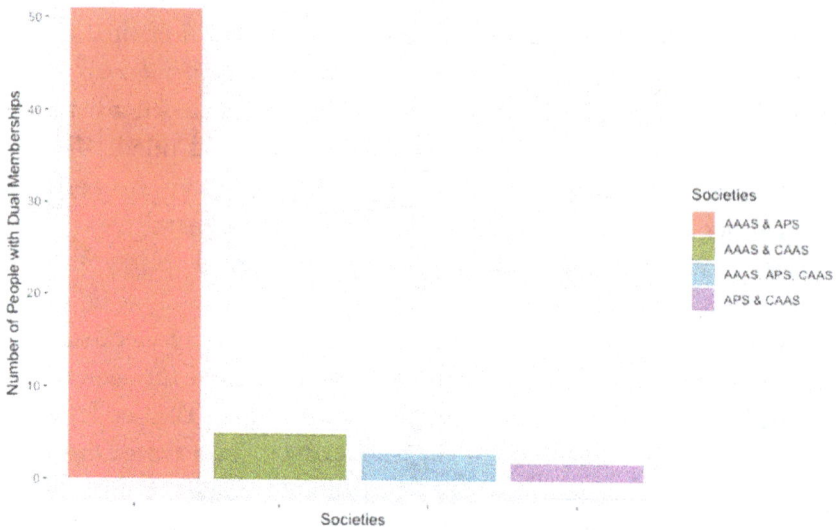

Figure 5.4 Number of overlapping memberships among the three societies.

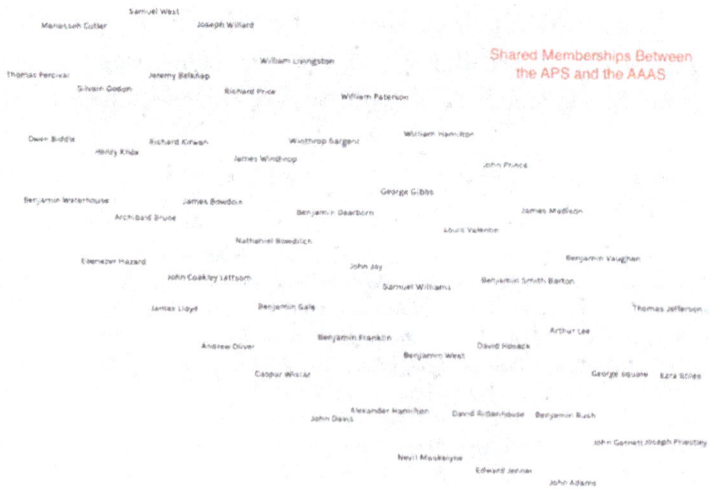

Figure 5.5 Shared memberships between the American Philosophical Society and the American Academy of Arts and Sciences.

Benjamin Silliman

David Humphreys   John Trumbull

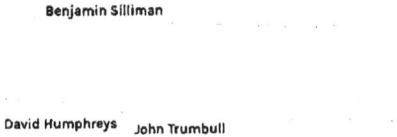

Figure 5.6 Shared memberships among the American Philosophical Society, the American Academy of Arts and Sciences, and the Connecticut Academy of Arts and Sciences.

This lack of shared memberships in these groups indicates that there may be political concerns among the various groups that transcend the scientific qualifications of a person.

The shared connections among memberships of all three societies were John Trumbull, David Humphreys, and Benjamin Silliman. John Trumbull and David Humphreys were part of a group of writers known as the *Connecticut Wits*. Humphreys, in particular, was associated with the Federalist Party and a political enemy of Jefferson and the Republicans. Benjamin Silliman was universally seen as a leading scientific figure. He considered the older institutions, such as the APS and the AAAS, as organizations "filled with amateurs and philosophes" instead of men with scientific training.[22] Perhaps this indicates Silliman's desire to separate science from politics.. Nevertheless, other intriguing connections are seen when we consider the shared memberships between the CAAS and the AAAS (Figure 5.7).

Interesting shared memberships occur in the CAAS and the AAAS: Timothy Dwight, Oliver Ellsworth, Noah Webster, Jonathan Trumbull, and

---

[22] Chandos Michael Brown, *Benjamin Silliman: A Life in the Young Republic* (Princeton, NJ: Princeton University Press, 1989), 227.

Jonathan Trumbull
Jeremiah Day
Timothy Dwight
Oliver Ellsworth
Noah Webster

Figure 5.7 Memberships shared between the American Academy of Arts and Sciences and the Connecticut Academy of Arts and Sciences.

Jerimiah Day. These figures were all leading critics of Jefferson and the Republican Party's political policies. They focused on the expansionist policies and natural history interests of Jefferson and the scientific institution promoting westward expansion, which was the APS. As president of the APS, Jefferson was particularly interested in promoting knowledge about the natural history of America's vast new territories acquired from France in 1803. The APS served as a scientific and educational institutional nexus for the western expeditions meant to explore North America. One historian explains, "the Society in those early years of the republic often served as national library, museum, and academy of sciences."[23] The Lewis and Clark expedition is well known, but two other contemporary major scientific exploratory expeditions ventured into unknown zones of the continent. The members of these expeditions often relied on their scientific training to do the work assigned to them, and the leading scientific experts in the United States were all members of the APS. It was critical for the membership to learn astronomy, mathematics, surveying, and advanced mapmaking from APS member Andrew Ellicott. Also, they learned natural history and preservation strategies for plants an animal remains from

---

[23] Edward Carlos Carter, *"One Grand Pursuit": A Brief History of the American Philosophical Society's First 250 Years, 1743–1993* (Philadelphia: American Philosophical Society, 1993).

Charles Willson Peale and Benjamin Smith Barton. The APS remained a prominent scientific institution continuing into the 1820s, supporting the work of Major Stephen Long's expedition of the Great Plains.[24]

The definition of what constituted valuable information or scientific knowledge remained controversial. Even as Jefferson encouraged such endeavors, Federalists made these expeditions a political target. Many scoffed that the trips to the West represented a waste of time and money. Even worse, many Federalists saw these schemes as a ploy hatched by mediocre men with designs on participating in intellectual discourse within the public sphere. Federalists like Timothy Dwight viewed these men as false scientists who were

> authors of vain and deceitful philosophy; of Science falsely so-called; always full of vanity in their discoveries: Scoffers walking after their own lusts, and alluring others, through the same lusts, to follow them; promising them liberty, as their reward, and yet being themselves, and making their disciples, the lowest and most wretched of all slaves, the slaves of corruption. Philosophical pride, and the love of sinning in security and peace, are, therefore, the two great causes of Infidelity, according to the scriptures.[25]

The dissociation with the Holy Scripture offended Federalists enough; however, that when adding the threat of Jeffersonian pandering to those deemed the uneducated citizenry, the Jeffersonians became intolerable. Several cases exemplify the animosity between Federalists and those associated with the APS.[26]

It is significant to note that even when the groups agreed they did not act in unison. In January 1800, the APS acknowledged receipt of the Constitution of the newly established CAAS, and its memorial seeking the "cooperation of the American Philosophical Society in prevailing on Congress to direct the Census to be taken in a more detailed manner."[27]

---

[24] McDonald and McDonald, "West from West Point: Thomas Jefferson's Military Academy and the 'Empire of Liberty'"; William Ragan Stanton, American Philosophical Society, and Library, *American Scientific Exploration, 1803–1860: Manuscripts in Four Philadelphia Libraries* (Philadelphia: American Philosophical Society Library, 1991); George D. Oberle, "Institutionalizing the Information Revolution: Debates over Knowledge Institutions in the Early American Republic" (PhD diss., George Mason University, 2016), 71–72.

[25] Timothy Dwight, *A Discourse on Some Events of the Last Century, Delivered in the Brick Church in New Haven, on Wednesday, January 7, 1801* (New Haven, CT: Ezra Read, 1801), 20–21.

[26] Linda K. Kerber, *Federalists in Dissent; Imagery and Ideology in Jeffersonian America* (Ithaca, NY: Cornell University Press, 1970); David Dzurec, "Of Salt Mountains, Prairie Dogs, and Horned Frogs," *Journal of the Early Republic* 35, no. 1 (Spring 2015): 79–108; Greene, *American Science in the Age of Jefferson*, 1984, 138.

[27] American Philosophical Society, "Early Proceedings of the American Philosophical Society for the Promotion of Useful Knowledge, Compiled by One of the Secretaries, from the Manuscript Minutes of Its Meetings from 1744–1838," *Proceedings of the American Philosophical Society* 22, no. 119 (July 1885), 290.

They especially hoped that the APS would support their request that the "the Census to be taken in a more detailed manner" to allow the "young and flourishing republic to become acquainted with its own natural history."[28] In December 1799 the AAAS also received the request from the CAAS to lobby Congress for a census that collected a variety of useful data on the "health and longevity of the Citizens of the United States."[29] The APS responded quickly with its own memorial written to Congress that called for a detailed statistical accounting of the inhabitants of the new country in order "to determine the effect of the soil and climate of the U. S. on the inhabitants thereof" and to "more exactly distinguishing the increase of population by birth and by immigration."[30] The APS further wanted the census to identify one of nine types of professions and vocations to which all free males belonged and "from these data truths will result very satisfactory to our citizens, that under the joint influence of soil, climate & occupation the duration of human life in this portion of the earth will be found at least equal to what it is in any other; and that its population increases with a rapidity unequaled in all others."[31]

This episode provides us with insight into a growing desire for increased collaboration among the learned societies, which served as a core institution designed to create and disseminate knowledge in the Early Republic and in connection with the government. Some scholars believe that Congress ignored these requests; however, they were taken seriously enough to be referred to the committee responsible for the census.[32] This call for the establishment of a national scientific priority by the leading intellectuals of the day ought to have been a natural addition to the census. However, the census remained fundamentally unchanged for several years. Why?

There were fundamental differences between the proposals of the CAAS and the APS, and the AAAS did not respond at all. The APS, for example, only asked to count the demographic information by age intervals of free males, whereas the CAAS looked to understand the entire population. The CAAS also wanted to know the number of married and unmarried

---

[28] Connecticut Academy of Arts and Sciences to American Philosophical Society, 25 December 1799. Archives, American Philosophical Society.

[29] Letter from Connecticut Academy of Arts and Sciences to the American Academy of Arts and Sciences, 24 December 1799, Series I-B-1: General records. Letterbooks. Bound letterbooks. Volume 2, 1792–1803. Archives, American Academy of Arts and Sciences, Cambridge, MA, P2-104, https://www.amacad.org/archive/images/001252.002.jpg.

[30] American Philosophical Society, "Early Proceedings of the American Philosophical Society," 293.

[31] American Philosophical Society, "Early Proceedings of the American Philosophical Society," 294.

[32] Margo J. Anderson, *The American Census: A Social History*, 2nd ed. (New Haven, CT: Yale University Press, 2015), 18–19.

people, whereas the APS preferred detailed occupational data. Scholars
have not explained the silence from the AAAS. In a letter from Joseph
McKeen to the AAAS, the minister and future president of Bowdoin College
believed that there were not enough "men of accuracy" who could be
employed for the gathering of the "multiplicity of objects" being called for
and that "innumerable errors would be the certain consequence."[33] McKeen
supported the idea being called for but feared that if done improperly, the
result would harm implementation of the overarching goals. Therefore, he
suggested establishing a list of priorities for new additions to the census.
No list of priorities was created. The AAAS never sent a response to
Congress, nor is there any surviving evidence of its reply to the CAAS or
the APS. The AAAS held the same desire to promote the collection of new
data, which it saw as knowledge. Political differences led to inaction in
Congress when it came to significant changes in the census, although some
of the suggestions were added to the fourth census in 1830.[34] In the
meantime, the CAAS conducted its own study of its state, collecting the
statistical data from townships in Connecticut that they hoped to collect
for the country. James Morris and Timothy Dwight published their results
in different editions of *A Statistical Account of the Towns and Parishes of
the State of Connecticut* as early as 1811. A few years later, in 1813,
Rodolphus Dickinson, not associated with any learned society, produced
the book *A Geographical and Statistical View of Massachusetts Proper*,
which mirrored other descriptive guides of the era and did not take the
systematic approach called for by the CAAS.

The differences between the contrasting visions of these projects
provide insights into epistemological differences between members of the
societies. Even though each of these groups was interested in collecting
data, they disagreed greatly as to what constituted useful data. These
patterns are more readily seen when combining traditional methods of
historical analysis with the ideas and basic techniques of digital scholarship.
This chapter demonstrates that these learned societies studied similar types
of knowledge, yet some within the CAAS dictated a research agenda that

[33] Letter from Joseph McKeen to Eliphalet Pearson, 15 January 1800. Series I-B-1: General records. Letterbooks. Bound letterbooks. Volume 2, 1792–1803. Archives, American Academy of Arts and Sciences, Cambridge, Massachusetts, P2-105 (https://www.amacad.org/archive/images/001253.001.jpg).
[34] Carroll Davidson, William C. Hunt, *The History and Growth of the United States Census*, (New York: Johnson Reprint Corp., 1966), 25–27; also see chapter 5. Patricia Cline Cohen, *A Calculating People: The spread of Numeracy in Early America*, (Chicago: University of Chicago Press, 1982).

opposed that of the Jeffersonian-dominated APS.[35] Further, using digital-history strategies offers hints for scholars to explore using the manuscript collections of the shared members.

The manuscript collections of the five members of the CAAS and the AAAS identified in Figure 5.7 provide useful insights into the way politics impacted the development of scientific knowledge in the Early Republic. There are opportunities to add more learned societies into this analysis if we extend our analytical period through the end of the so-called Era of Good Feelings.[36]

---

[35] This chapter attempts to use what Lincoln Mullen has called a "braided narrative for digital history." This approach argues that digital scholarship needs to weave our explanation of methods, which is most common among digital scholarship, into our narrative, which is focused on making an argument. See Lincoln Mullen, "A Braided Narrative for Digital History," in *Debates in the Digital Humanities 2019*, eds. Matthew K. Gold and Lauren F. Klein (Minneapolis: University of Minnesota Press), 2018.

[36] I am very grateful to Alyssa Fahringer, Digital Scholarship Consultant, who was very helpful with using ggplot in Rstudio; as well Cynthia Kierner and Mark Boonshoft, who provided exceptional feedback on this essay. Also, the librarians and archivists of the American Philosophical Society have provided many hours of assistance for which I am eternally grateful. I, of course, take full responsibility for the final content.

# Index

Page numbers in *italics* refer to illustrations.

www.ingramcontent.com/pod-product-compliance
Lightning Source LLC
Chambersburg PA
CBHW081415160426
42812CB00087B/2332